THE INTERPRETATION OF QUANTUM MECHANICS

Erwin Schrödinger

THE INTERPRETATION
of
QUANTUM MECHANICS

Dublin seminars (1949–1955) and other
unpublished essays

Edited and with Introduction by Michel Bitbol

Ox Bow Press
Woodbridge, Connecticut 06525

Published by
OX BOW PRESS
P.O. Box 4045
Woodbridge, Connecticut 06525
203 387-5900 (voice)
203 387-0035 (fax)

Copyright © 1995 by Ruth Braunizer

Printed in the United States of America.

Library of Congress Cataloging-in-Publication Data
Schrödinger, Erwin, 1887–1961
    The interpretation of quantum mechanics : Dublin seminars (1949–1955) and other unpublished essays / Erwin Schrödinger ; with introduction and notes by M. Bitbol.
    p. cm.
    Includes bibliographical references (p.).
    ISBN 1-881987-08-6. — ISBN 1-881987-09-4 (pbk.)
    1. Quantum theory. I. Title.
QC174.12.S35   1995
530.1'2—dc20   95-18611
                                                                        CIP

The paper in this book meets the guidelines for permanence and durability of the Committee on Production Guidelines for Book Longevity of the Council on Library Resources.

Cover photograph by Prof. Dr. Wolfgang Pfaundler.

Contents

• INTRODUCTION • 1

1. Interpretations or Interpretation? 1
2. The Fading of the Concept of Particle 5
3. An Ontology of Ψ-Waves 9
4. The Measurement Problem 13
5. Sources 18
   Typographic Conventions 18

1 • JULY 1952 COLLOQUIUM • 19

1. Introduction 19
2. How It Came About 22
3. The Alleged Energy Balance—A Resonance Phenomenon 24
4. Removing Some Difficulties 26
5. Chemistry, Photochemistry, and the Photoelectric Effect 28
6. (Individuality and Sameness) 31
7. Thermodynamics 36

2 • TRANSFORMATION AND INTERPRETATION IN QUANTUM MECHANICS (C. 1952) • 39

Introductory Remark 39
1. The Wave Function; Linear Operators; Eigenfunctions and Eigenvalues 40
2. The Transformation to Another Frame 44

　　　　3　The Transformation of the Wave Function　46
　　　　4　The Interpretation of the Wave Function　49
　　　　5　Illustrations　54
　　　　6　(Harmonic Oscillators)　63
　　　　7　(Interpreting the Wave Function)　70
　　　　8　(Laws and Changes with Time)　82

3 • NOTES FOR 1949 SEMINAR • 97

　　　　The Problem of Matter in Quantum Mechanics　97
　　　　The Nature of the Elementary Particles　99
　　　　The Difficulties in Interpreting the Blur (Quantum Uncertainty)　104

4 • NOTES FOR 1955 SEMINAR • 109

　　　　Introduction　109
　　　　On Measuring Velocities by the Police or Race-Course Method　110
　　　　Introduction to Statistics　114
　　　　Planck-Black-Body-Radiation (Without Discontinuity!)　115

5 • FROM A LETTER TO ARTHUR S. EDDINGTON, MARCH 22ND 1940 • 121

6 • WILLIAM JAMES LECTURES • 123

　　　　1　First Lecture: Science, Philosophy and the Sensates　123
　　　　2　Second Lecture: The Technique of Measurement　131
　　　　3　Third Lecture: The Part of the Human Mind　141

• ABOUT THE BOOK • 151

• Introduction

Erwin Schrödinger[1] apparently changed his mind four times about the interpretation of quantum mechanics during the quarter of a century after the birth of the theory. In order to put the writings of Schrödinger published in the present volume in their intellectual context, it is necessary to state these four interpretations, and then to clarify their nature. As we shall see, these interpretations can be construed as successive steps in a single project of ontological reconstruction rather than as completely distinct frameworks of thought. According to this analysis, the texts of the early 1950s, and especially the Dublin seminars of that period, are to be regarded as Schrödinger's most elaborate and lucid attempt at expounding his project.

1 • Interpretations or Interpretation?

Schrödinger's first interpretation of quantum mechanics was sketched in January and February of 1926, in the pioneering papers entitled "Quantization as a problem of proper values" (I and II).[2] It amounted to taking the ψ-function at face-value and considering it as a direct description of standing wave-like processes occurring within the boundary of atoms. However, in the early spring of 1926, Schrödinger

---

[1] Further developments on Schrödinger's interpretation(s) of quantum mechanics and general philosophy can be found in: M. Bitbol, *Schrödinger's philosophy of quantum mechanics*, Kluwer, 1996 (to be published); M. Bitbol & O. Darrigol (eds.), *Erwin Schrödinger, Philosophy and the birth of quantum mechanics*, Editions Frontières (1992); M. Bitbol, *L'élision*, introductory essay to: E. Schrödinger, *L'esprit et la matière*, Seuil, 1990.

[2] E. Schrödinger, "Quantization as a problem of proper values" (I and II), in: *Collected papers on wave mechanics*, Blackie and son, 1928.

realized that this way of dealing with the wave-mechanical formalism could only lead one to the proper modes of the vibrating system; it provided no clue for a satisfactory account of line intensities and polarizations. As soon as he had demonstrated the mathematical equivalence of his wave mechanics with Heisenberg's, Born's, and Jordan's matrix mechanics, Schrödinger concluded that the ψ-function had to be taken as an intermediate-level concept *(hilfbegriff)*, and that the correct description of the atomic processes was actually afforded by the product $-e\psi\psi^*$, considered as an electric charge density.[3] This second approach was only partially successful, however, and it did not remove all the difficulties that plagued the original wave interpretation. Schrödinger's initial attempts came to an abrupt end with the onset of Born's probabilistic interpretation of ψ, the pressure of strong criticism from the Göttingen-Copenhagen physicists, and their elaboration of a full-blown synthetic apprehension of quantum mechanics whose two cornerstones were Heisenberg's "uncertainty relations" and Bohr's complementarity principle. From 1928 on, Schrödinger decided to teach quantum mechanics according to the mainstream "Copenhagen interpretation." We can consider this to be his third interpretation of quantum mechanics, even though it was one that he borrowed from the Göttingen-Copenhagen group. The year 1935 marked a noticeable change. A few weeks after the appearance of the Einstein-Podolsky-Rosen paper and a rich subsequent correspondence with Einstein, Schrödinger published both his "cat-paper"[4] and a more technical article about the "entanglement" of wave-functions.[5] In these two pieces of work, Schrödinger expressed a well-documented skepticism about the current interpretation of quantum mechanics, even though he was admittedly unable to offer any satisfactory alternative. Finally, in the late forties and early fifties, he became increasingly self-assertive, and clung to a personal conception of quantum mechanics.

---

[3] See L. Wessels, *Schrödinger's interpretations of quantum mechanics*, Indiana University, Ph.D., 1975.

[4] The two papers are: A. Einstein, B. Podolsky, and N. Rosen, "Can quantum-mechanical description of physical reality be considered complete?" *Phys. Rev.* 47, 777–780, 1935; E. Schrödinger, "Die gegenwärtige Situation in der Quantenmechanik," *Naturwissenschaften*, 23, 807–812, 823–828, 844–849, 1935. They are both reprinted, in English translation, in: J. A. Wheeler and W. K. Zurek, *Quantum mechanics and measurement*, Princeton University Press, 1983.

[5] E. Schrödinger, "Discussion of probability relations between separated systems," *Proc. Cambridge Phil. Soc.*, 31, 555–563, 1935.

Many of his colleagues[6] assumed that it was merely a revival of his first 1926 wave-interpretation.

A closer analysis, however, shows that this account of Schrödinger's successive interpretations of quantum mechanics is superficial, overlooking two significant points.

First, very soon after he proposed his second (or electrodynamic) interpretation, Schrödinger became fully aware that it could not work alone, that somehow it had to be combined with the original wave interpretation. Whereas the electrodynamic interpretation enabled one to make sense of the line intensities and polarizations, the wave representation was still needed to calculate the relevant charge distributions as well as the evolution between two measurements of line intensities.[7] The problem Schrödinger had to face was the following: we need not one but *two* representations of the atomic phenomena. One of them, namely the electrodynamic representation, is directly related with the "observed facts," but it does not provide the link *between* subsequent observed facts; in a word, it is "factual" but not "effective." The second one, namely the wave representation, is perfectly able to provide a link between the observed facts, but not to account for all the aspects of the facts themselves; it is "effective" but not "factual." As long as Schrödinger wanted to merge the "effective" and the "factual" into a single representation, according to the classical ideal, the persistent duality of the models had to be considered as a symptom of failure.

But in the 1950s, he accepted the complete dissociation of the representation (namely the wave picture) from the facts and the natural processes underlying them: "We do give a complete description, continuous in space and time without leaving any gaps, conforming to the classical ideal—a description of something. But we do not claim that this 'something' is the observed or observable facts; and still less do we claim that we thus describe what nature (. . .) really

---

[6] See, e.g., M. Born, "The interpretation of quantum mechanics," *Brit. J. Phil. Sci.* 4, 95–106, 1953.

[7] A typical example of this combined use of the wave and electrodynamic interpretations can be found in: E. Schrödinger, "The Compton effect," *Ann. der Phys.*, 82, 1927; in: *Collected papers on quantum mechanics*, op. cit. p. 124. For a valuable comment, see J. Dorling, "Schrödinger's original interpretation of the Schrödinger equation: a rescue attempt," in: C. W. Kilmister (ed.), *Schrödinger, centenary celebration of a polymath*, Cambridge University Press, 1987.

is."[8] Schrödinger's aim was thus no longer to unify his two original representations but rather to understand the necessity of using both. One of them (namely the "effective" wave representation) accordingly gained a privileged status, whereas the other one (the "factual" electrodynamic representation, which is formally equivalent to the Born probabilistic rules) was considered as nothing more than a formulation of the link between the wave representation and the observed facts.

It is now clear that Schrödinger's two initial interpretations of quantum mechanics were not two completely distinct conceptions, but that from the outset they had to be used jointly. As for Schrödinger's last interpretation of the 1950s, it appears that it was not a mere rehearsal of the first wave interpretation, but rather a self-conscious and systematic application of the idea that, in general, a representation does not mimic directly the experimental facts; that its connection with facts rather occurs through highly non-trivial empirical correspondence rules.

Second, it is usually stated that Schrödinger's successive attempts at giving a satisfactory interpretation of quantum mechanics are associated with a kind of epistemological wavering, and that his first and last interpretations arise from a realist attitude; whereas his intermediate acceptance of the Copenhagen interpretation is strongly linked with an anti-realist outlook. Indeed, one can extract from the 1926 and 1950 papers some sentences which sound like an expression of a very strong realist commitment, whereas some conferences during the beginning of the 1930s are sometimes more extremist in their anti-realist formulation than the texts of the most extremist members of the Copenhagen group. It is, however, quite easy to make sense of these apparent changes if one remembers Schrödinger's metaphysical background. He called the kind of view he favored an "idealistic monism,"[9] inspired by Mach's doctrine of sensations-elements, by Russell's *Analysis of Mind*, and by the post-Kantian philosophy of early nineteenth-century Germany. His insistence on the reality of theoretical entities such as the wave-function was therefore not dependant on a metaphysical realism, but rather on a methodological scientific realism which involved a strong belief in the *intellectual* value of models and representations in science.[10] Saying that the wave func-

---

[8] E. Schrödinger, *Science and Humanism*, Cambridge University Press, 1951, p. 40.

[9] E. Schrödinger, *My View of the World*, Ox Bow Press, 1983, II-1.

[10] See this volume, p. 121, about the influence Boltzmann exerted on his thought together with Mach.

tions are "real" was thus by no means equivalent to ascribing them the status of *Ding-an-sich*. It was rather a way of underlining that, as theoretical constructs, they are not inferior to the constructs that we had to form during our childhood in order to survive in our environment, namely the material bodies of our everyday life.[11] The anti-realist step of the 1930s then appears in a very different light. Rather than a temporary renunciation of realism, it can be viewed as an intermediary period of analysis of the data that could be used in the future to *reconstruct* something out of it; that is, used to reconstruct entities which one would have good reasons to consider exactly *as if* they were "real."

At the heart of Schrödinger's late reflections on the interpretation of quantum mechanics, therefore, lies the question whether one can form a new ontology in strict correspondence with the symbolic system of standard quantum mechanics.

The preliminary step in this direction consisted in dismantling more carefully than ever the traditional ontology of localized bodies, and in seeing whether its elements could be transferred as they stand to a novel system of entities. In this light, we shall state Schrödinger's attitude toward the concept of particle and then analyze in detail the arguments Schrödinger gave in favor of his new ontology of Ψ-waves. Thus we shall see that the main difficulty that hindered Schrödinger's undertaking was the lack of a proper solution to the measurement problem. We will further explore Schrödinger's original method of addressing, if not solving, the measurement problem.

2 • The Fading of the Concept of Particle

What is a particle? It is a small localized body with three constitutive features, none of which are necessarily independent of one another:

(i) It can be ascribed a set of *properties* which embody *virtual* observations expressed by *counterfactual* empirical propositions.

(ii) It has individuality.

(iii) It can be re-identified through time.

Our strategy of exploration must consist of studying the three listed ontological features while mentioning, whenever necessary, their deep-lying inter-relations.

---

[11] See this volume, p. 148.

The concept of virtuality was very soon recognized as a cornerstone in the debate on the interpretation of quantum mechanics. As early as 1926, Einstein challenged Heisenberg's positivist-like strict adherence to *actual* results of experiments. He believed that one could not dispense with introducing some version of the modal category of the *possible* in the reasonings, and retain just the *actual*,[12] lest one lose the very content of the notion of a real object on which experiments are performed. In a conference of 1928, Schrödinger went even further than Einstein by stressing the decisive importance of *virtualities* as a basic ontological constituent. Whereas Einstein considered the "virtual" or the "foreseeable" as a component of reality, Schrödinger *defined* reality as a construct made out of a proper combination of actual and virtual material: "That is the *reality* which surrounds us: some actual perceptions and sensations become automatically supplemented by a number of virtual perceptions and appear connected in independent complexes, which we call existing objects."[13] This sentence, together with other similar ones, defines Schrödinger's peculiar use of the concept of virtualities. Firstly, the "virtual" perceptions, observations, or experimental results, which constitute a real object, are associated in "complexes"; they are construed as co-existent; in short, they are listed in terms of *conjunction* rather than *disjunction*. Secondly, the justification of their being linked in such a way is that they are experimentally *accessible* at any moment. Virtualities are conceived by Schrödinger as a modal expression of expectations: "We are not usually aware of all these expectations; we focus them unconsciously into what we call a fruit basket which really exists."[14]

Of course, one has to qualify this condition of permanent accessibility to the virtual observations in order to make it applicable to most familiar situations of daily life. An ideal accessibility presupposes that no changes whatsoever happen between the instant when the actual observation is made and the instant when the conditions of the expected observation are fulfilled. But generally this is not true. Some disturbances may occur, or the system may be subject to a law of evolution which modifies its state in the interval. It is thus indispensable

---

[12] W. Heisenberg, *Physics and beyond, encounters and conversations*, George Allen & Unwin, 1971, Chapter 5.

[13] E. Schrödinger, "Conceptual models in physics and their philosophical value," in: *Science, theory and man*, Dover, 1957.

[14] Ibid. p. 120.

to *modulate* the condition of accessibility by a certain relevant evolution factor.

Faced with the fact that any pair of measurements of two conjugate variables such as position and momentum is submitted to Heisenberg's uncertainty relations, Schrödinger had found good reason to be pessimistic about "(. . .) whether in this case, in principle, virtual observations are at all conceivable, on which the real existence of these objects can be based."[15] True, the idea that particles *have* underlying properties of the usual sort, but that each one property is disturbed[16] in an *uncontrollable* way by the measurement of the other conjugate one, could still be sustained at this early stage of the debate about the meaning of quantum mechanics; such a possibility would have been sufficient to maintain, at least formally, the concept of virtuality in spite of the uncertainty relations. But Schrödinger found it increasingly difficult to accept. He had formulated his own version of the von Neumann no-hidden-variable theorems in his 1935 "Cat's paper," and he thus underlined repeatedly in the 1950s that the idea that particles possess values of every observable is just a "belief."[17] When applied to the observables **q** and **p**, these remarks led him to the conclusion that the particles cannot even be ascribed anything like a continuous trajectory: "Observations are to be regarded as discrete, disconnected events. Between them there are gaps which we cannot fill in."[18]

It is at this point that Schrödinger's hyper-revolutionary attitude arises. Is it coherent to keep on speaking of "particles" if they have nothing like a trajectory? Schrödinger's answer is a definite *no*. When he asked "what is a particle which has no trajectory or no path?"[19], it was just a somewhat ironical way of emphasizing that "(. . .) the particles, in the naive sense of the old days, *do not exist.*"[20] Some years later, he confirmed most clearly this equivalence between

---

[15] Ibid. p. 121.

[16] For a criticism of the disturbance conception of measurements, see E. Schrödinger, "What is an elementary particle?" *Endeavour*, XI, 110–116, p. 111.

[17] See this volume, p. 79.

[18] E. Schrödinger, *Science and Humanism*, Cambridge University Press, 1951, p. 27.

[19] E. Schrödinger, "L'image actuelle de la matière" in: *Gesammelte abhandlungen*, Friedr. Wievweg & Sohn, 1984, vol. 4, p. 507.

[20] Ibid. p. 506.

no trajectory and *no particle at all* in a letter to Henry Margenau: "To me, giving up the path seems giving up the particle."[21]

The reason for this strict implication is to be found in Schrödinger's meditation about individuality. The "individual sameness" of the macroscopic bodies which surround us is ascertained, according to him, by their "*form* or *shape* (German: *Gestalt*),"[22] including some imperceptible details which distinguish them from other bodies of the same kind. The elementary particles can also be ascribed a form in the broad sense of set of actual properties. But this form can only define their *kind*; it is not sufficient in general to single out each one of them. Thus one has to revert to another criterion in order to ascertain the individuality of the particles. The alternative criterion is simply based on their having distinct past histories or distinct trajectories. As Schrödinger himself noticed in his letter to Margenau, this criterion was already widely recognized in the time of classical mechanics, and it was explicitly proposed by Boltzmann in his *Vorlesungen über die Principe der Mechanik* of 1897.[23] But Schrödinger stressed that, due to the uncertainty relations (supplemented with his version of no-hidden-variable theorem), the particles, if any, *cannot* be ascribed a well-defined trajectory. The ultimate criterion of individuality thus collapses.

Of course, one possibility of rescuing the old concept of individual body remained. It was to say that two groups of circumstances are to be distinguished: the circumstances where the range of uncertainty of two trajectories overlap on the one hand, and the circumstances where they do not overlap on the other hand. In the first case, the particles have no definite individuality, whereas in the second case, they have one. But Schrödinger rejected this expedient. According to him, "(. . .) there is no sharp boundary, no clear-cut distinction between (the two types of circumstances), there is a gradual transition over intermediate cases."[24] Even if two "particles" are experimentally located very far apart from each other, even if their Δx's do not overlap, there is still a small probability that an "exchange" has occurred between them. The distinction can thus be performed *in practice*, but

---

[21] E. Schrödinger to H. Margenau, April 12, 1955, AHQP, microfilm 37, section 9.

[22] E. Schrödinger, *Science and Humanism*, op. cit. p. 19.

[23] See: L. Boltzmann, *Theoretical physics and philosophical problems* (B. Mac Guinness, ed.), Reidel, 1974, p. 230–231.

[24] E. Schrödinger, *Science and Humanism*, op. cit. p. 17.

its possibility is ruled out *in principle:* "I beg to emphasize this and I beg to believe it: It is not a question of our being able to ascertain the identity in some instances and not being able to do so in others. It is beyond doubt that the question of 'sameness,' of identity, really and truly has no meaning."[25] In principle, there is nothing like two *distinct particles*. There is thus nothing like an individual and transtemporally reidentifiable particle; and, Schrödinger concludes, there is thus nothing like a particle, only chains of instantaneous events.[26]

3 • An Ontology of Ψ-Waves

Insofar as the previous analysis has left one with only scattered events, the question of their lawlike connection arises. Admittedly, no pure corpuscularian representation can afford the sought-after connection. This is so because, even if the concept of trajectory could be maintained, it would only provide one with a *longitudinal* linkage between the events, whereas quantum phenomena also display a *transversal* linkage which manifests itself through the interference patterns.[27] Are we then compelled to adopt something like Bohrian complementarity, between a symbolic particle picture expressing the longitudinal linkage of events, and a symbolic wave picture expressing their transversal linkage? Schrödinger did not think so. Waves can do both jobs at once. Indeed, while the concept of particle path only bears longitudinal linkage, the concept of (multi-dimensional) wave synthesizes the *two* types of linkages: "In a wave phenomenon you have—not always, but in many cases—the two complementary features of wave- (or phase-) surfaces and of wave-normals or rays."[28]

The multi-dimensional wave-functions, however, remained up to this point abstract entities, embodying the quantum mechanical lawlike connection between otherwise isolated experimental events. But Schrödinger found that they have at least four characteristics which support their being ontologically construed (at least if one takes

---

[25] Ibid. p. 18.

[26] E. Schrödinger, "What is an elementary particle?" loc. cit.

[27] E. Schrödinger, "L'image actuelle de la matière" in: *Gesammelte abhandlungen*, op. cit., vol. 4, p. 506.

[28] See this volume, p. 20.

"ontology" in its modern semantical acceptation rather than with its traditional metaphysical connotations[29]).

(i) Wave-functions are so defined that they do not share the major defect of the corpuscularian representation, namely that the latter "(. . .) constantly drives our mind to ask for information which has obviously no significance."[30] The corpuscularian representation drives our mind to ask for the precise value of the momentum of a particle at the very instant when the position observable has been measured; and the fact one cannot answer this question is ascribed to the measuring devices' being mutually incompatible. By contrast, a wave-function is perfectly defined when only *one* of the two canonically conjugated observables (position *or* momentum) has been ascribed a precise value. No mention of the incompatibility of our instruments is required.

In order to avoid being compelled to adopt a reflexive or epistemological attitude, one has to make sure that the newly defined ontology does not leave any occasion to ask questions which are experimentally unanswerable. It must reflect so perfectly the available experimental circumstances that it becomes superfluous to refer explicitly to them: "An adequate picture must not trouble us with this disquieting urge; it must be incapable of picturing more than there is; it must refuse any further addition."[31]

(ii) The "effectiveness" of the wave functions, namely their ability to embody the lawlike connection between experimental events was considered by Schrödinger as a good reason for calling them "real," even if it is in a very prudent sense of the word *real* that is much more akin to Simon Blackburn's quasi-realism[32] than to any metaphysical version of realism. Commenting on de Broglie's concept of "guiding wave," he noted:

---

[29] Schrödinger explicitly rejected the project of "framing ontologically" the elements of our physical pictures, if "ontologically" is taken *in the metaphysical sense*, see, e.g., "Might perhaps energy be a merely statistical concept?" in: *Gesammelte abhandlungen*, vol. 1, p. 508.

[30] E. Schrödinger, "What is an elementary particle?" loc. cit. p. 111.

[31] Ibid.

[32] Simon Blackburn, *Essays in quasi-realism*, Oxford University Press; M. Bitbol, "Quasi-réalisme et pensée physique," *Critique*, 564, 340–361, 1994; M. Bitbol, *Schrödinger's philosophy of quantum mechanics*, Kluwer 1996 (to be published).

"Something that influences the physical behaviour of something else must not in any respect be called less real than the something it influences—whatever meaning we give to the dangerous epithet 'real'."[33]

This feature of lawlike ordering was a much more momentous feature of phenomena, according to Schrödinger, than the occurrence of isolated experimental events: "We are interested in general laws, not in special facts."[34]

Our ontology must then rely exclusively on lawlike connections, rather than on composite aggregates of experimental events. An ontology of wave functions is fully consistent with this condition, since what the quantum mechanical laws rule is just the evolution of these wave functions in time.

(iii) The ψ-function of quantum mechanics directly indicates "simultaneous happenings" on wave-surfaces, rather than "alternatives"[35]; it is this circumstance which gives rise to interference patterns. Consequently, the said (virtual) "happenings" are to be listed in the form of a *conjunction* rather than in the form of a *disjunction*. Schrödinger noticed very early that the use of conjunctions of coexistent (virtual) happenings was the crucial feature that distinguishes wave mechanics from particle mechanics: "We are confronted with the profound logical antithesis between

Either this or that (aut-aut)	(particle mechanics)
and	
This as well as that (et-et)."[36]	(wave mechanics)

Later on, in 1952, he underlined the ontological significance of these conjunctions, quite consistently with his former definition of real objects as constructs made of simultaneous occurrences: "Here 'real' is not a controversial philosophical term. It means that the wave acts simultaneously throughout

---

[33] E. Schrödinger, "What is an elementary particle?" loc. cit.
[34] See this volume, p. 81.
[35] See this volume, p. 20.
[36] E. Schrödinger, "The fundamental idea of wave mechanics" (Nobel lecture, 1933), in: *Science and the human temperament*, G. Allen & Unwin, 1935.

the whole region it covers, not either here or there. (. . .) So the epithet 'real' means the momentous difference between 'both-and' *(et-et)* and 'either or' *(aut-aut)*."[37]

There was a very important difficulty, however, which Schrödinger fully recognized. Whereas his conception of reality involved an aggregate of simultaneously occurring virtual and *actual* happenings, the actual happenings of quantum mechanics could not be treated on the same footing as the virtual ones. The wave formalism articulated conjunctions of (virtual) happenings, but on the other hand, whenever actual facts are concerned, there is no way by which one can avoid making use of disjunctions. As he noticed in 1935, during the measurement process, "The expectation-catalog of the object has split into a conditional disjunction of expectation-catalogs."[38] This difficulty is intricate indeed, and it is obviously related to the measurement problem. Schrödinger did not really tackle it; he contented himself with proclaiming again the priority of general laws over particular facts: "If you accept the current probability views *(aut-aut)* in quantum mechanics, the single event observation becomes comparatively easy to tackle, but all the rest of physics (. . .) is lost to sight."[39]

(iv) The ψ-waves are individuals. They are individuals in virtue of their having a form, namely a (frequency or amplitude) modulation.[40] There is, however, a big difference between these genuinely quantum mechanical individuals and the corpuscularian individuals of classical mechanics. Classical particles are individualized by the position they occupy in ordinary space at each point in time, namely by their trajectory, whereas the ψ-waves are individualized by their form in configuration space. An interesting particular case is that of standing waves, which are completely ubiquitous in the volume of configuration space they occupy, and yet have individuality in virtue of their form.[41]

---

[37] E. Schrödinger, "Are there quantum jumps?" *Brit. J. Philos. Sci.* 3, 1952, p. 242.

[38] E. Schrödinger, "The present situation in quantum mechanics," in: J. A. Wheeler and W. H. Zurek (eds.), *Quantum theory and measurement*, Princeton University Press, 1983, p. 159.

[39] E. Schrödinger, "Are there quantum jumps?" loc. cit. p. 242.

[40] E. Schrödinger, "L'image actuelle de la matière" in: *Gesammelte abhandlungen*, vol. 4; also this volume, p. 32.

[41] See this volume, p. 32.

To summarize: ψ-waves are able to avoid having recourse *explicitly* to epistemological considerations in the formulation of quantum mechanics; they are ruled by the law of evolution of this theory; they bear coexistent virtualities; and they are reidentifiable individuals. According to Schrödinger, these circumstances support ascribing ψ-waves the status of entities of a new ontology.

As we mentioned previously, the only weakness of this approach is related to the measurement problem: the actual happenings cannot be united to the virtual ones in a single conjunction of coexistent occurrences; accordingly, it cannot be said that they inhere in a single "real" entity.

Therefore, we must now attend to the measurement problem.

4 • The Measurement Problem

In agreement with Schrödinger's decision to dismiss any concession to the ontology of localized bodies, the experimental discontinuities could not reflect any corpuscular aspect of the microscopic processes; they had to arise from a peculiar feature of the interaction between the (wave-like) system and the (wave-like) apparatus: "One must regard the 'observation of an electron' as an event that occurs within a train of de Broglie's waves when a contraption is interposed in it which by its very nature cannot but answer by discrete responses."[42] Obviously, this sentence was by no means intended as a solution of the difficulty, but as a statement of the result which should *eventually* be reached. And Schrödinger was fully aware of the long distance which had to be covered before one can come to a satisfactory outlet in the direction he indicated. In particular, he thought that no metaphorical account of the transition from the continuous description to the experimental discontinuities, such as the "reduction of the wave packet" (or "wave packet collapse") initially suggested by Heisenberg,[43] could prove acceptable. But is there any alternative left? Schrödinger's arguments against the concept of "wave packet collapse" may at least help us outline, by contrast, the most likely features of this sought-after alternative.

---

[42] E. Schrödinger, "The meaning of wave mechanics," in: A. George (ed.), *Louis de Broglie physicien et penseur*, Albin Michel, 1953, p. 26.

[43] W. Heisenberg, "The physical content of quantum kinematics and dynamics" (1927), in: J. A. Wheeler and W. H. Zurek (eds.), *Quantum theory and measurement*, op. cit. p. 74.

In the 1950s, Schrödinger stated most clearly his reluctance to accept the "reduction of the wave packet" among the elements of the physical description. His position relied on a requirement of internal coherence of the theory, and of unity in its law of evolution: "If one accepts this law—and it is universally accepted as a general law—one must stick to it. It must not be occasionally infringed upon by a man making a measurement."[44]

As a consequence of this choice, he insisted on formulating the probabilistic correspondance rules[45] in such a way that they automatically rule out the idea of "collapse." This proved perfectly possible. It is indeed well known that one can choose between two (formally) equivalent formulations of the correspondence rules.[46] The first one requires an expression for the expectation value of an observable, whereas the second one (namely the Born's rule) is founded on a statistical axiom in the diagonal frame of the observable. Now, according to Schrödinger, even though the two formulations are formally equivalent, they are not *physically* equivalent: "(T)he second is shorter, but decidedly more artificial. You have to swallow a greater lump at a time. You have to assume explicitly that the system can never be found in a non-eigenstate, when this quantity is measured!"[47] The first formulation does not share this defect, for it does not incorporate any mention of the eigenstates.

In the same spirit, Schrödinger remarked that a pure wave-mechanical description of the interaction between the system and a measuring apparatus does not yield a "collapse" of the system's wave function, but rather its *entanglement* with the wave function of the apparatus. Following the norms of this type of description, the wave function of the system *disappears* in the melting pot of the wave function of the whole. True, one can perfectly use the information provided by the *actual outcome* of the measurement in order to extract a *new* wave function *for the system alone* out of the compound wave function. But this is by no means *a change* of the initial wave function of the system; this is a *redefinition* of it; this is a renewed *decision*

---

[44] See this volume, p. 83; see also E. Schrödinger, "The meaning of wave mechanics," in: A. George (ed.), *Louis de Broglie physicien et penseur*, op. cit., p. 18; and also this volume, p. 35).

[45] Or the "interpretation," in the restricted sense Schrödinger ascribed to this word: see this volume, p. 52.

[46] See this volume, p. 53.

[47] Ibid.

to separate the elements of information which had been entangled by the measuring process: "(. . .) it would not be quite right to say that the ψ-function of the object which changes *otherwise* according to a partial differential equation, independent of the observer, should *now* change leap-fashion because of a mental act. For it had disappeared; it was no more. Whatever is not, no more can it change. It is born anew, is reconstituted, is separated out from the entangled knowledge that one has (. . .)."[48] One can perfectly well *reconstitute* a system's wave-function by a "mental act," in order to predict as economically as possible the outcomes of subsequent measurements performed *on this system*, but it would be a *category mistake* in G. Ryle's sense to *mix* this choice (or "mental act") with the objective description: "(They say) one must not call it a physical change, it is only a change in our knowledge. I consider this an unfair subterfuge—or plainly: nonsense."[49]

Hence Schrödinger's general attitude towards quantum mechanics: push the description of its entities to its ultimate consequences; don't bother about their connection with experimental outcomes until the *very last* stage of the description; postpone the necessity of making the connection explicit as far as you can. "(O)ne must, to repeat this, hold on to the wave aspect throughout."[50] At most, one may rely on some loose stopping criteria which are sufficient *for all practical purposes*: "quantum mechanics stops as soon as anything reaches your senses (that has been said by Schopenhauer long ago)."[51]

As a consequence Schrödinger appeared as one of the few quantum physicists who felt motivated by formulating a genuine quantum theory of measurement. Actually, he was the physicist who made the very first step toward the modern quantum theory of measurement, as early as 1927.[52] The immediate purpose of this 1927 paper was to show that "without quantum postulates" one can arrive "at an effect which is exactly the same as if the quantum postulates was in force."[53]

---

[48] E. Schrödinger, "The present situation in quantum mechanics" in: J. A. Wheeler and W. H. Zurek (eds.), *Quantum theory and measurement*, op. cit. p. 162.

[49] See this volume, p. 83.

[50] E. Schrödinger, "The meaning of wave mechanics," in: A. George (ed.), *Louis de Broglie physicien et penseur*, op. cit. p. 26.

[51] See this volume, p. 97, particularly the footnotes.

[52] E. Schrödinger, "The exchange of energy according to wave mechanics," in: *Collected papers on wave mechanics*, op. cit.

[53] Ibid.

To obtain this result, Schrödinger gave a solution of his equation for the two interacting systems together, "united into *one* system." However, the general solution of the compound equation was not a simple product of the proper wave functions for each system, but their linear combination. In modern terms (and supposing one of the two systems is a measuring apparatus), one would say that his calculation enabled him to identify the *correlation* of the two systems through the *entanglement* of their wave functions, but not to identify the actual result of the measurement. Wave mechanics provided him with a discrete *scheme* of levels and processes, but it definitely had nothing to say about the singularity of each observed discrete phenomenon.

The measurement problem had not been *solved* by Schrödinger's paper of 1927. However, this paper was the first clear statement of it.

The measurement problem was stated even more clearly in 1935, when Schrödinger explicitly evoked the threat of infinite regress, whose seed was already present in the paper of 1927: "(. . .) this procedure will be called the *disentanglement*. Its sinister importance is due to its being involved in every measuring process and therefore forming the basis of the quantum theory of measurement, threatening us thereby with at least a *regressus ad infinitum*, since it will be noticed that the procedure itself involves measurement."[54]

In the 1950s, however, Schrödinger decided to resume his studies on the quantum theory of measurement. He had renewed reasons to do so. The Born rule, which Schrödinger was analyzing more seriously than ever before, required a rationale. But, since 1935, the quantum theory of measurement had not progressed very much. In 1963, when Margenau listed the major advances in this field,[55] he could but quote Von Neumann (1932), Schrödinger (1935), London and Bauer (1939), and himself (1936–1938). Schrödinger thus felt very lonely in his late undertaking.[56] He made some isolated developments in this direction, especially in his 1952 lectures entitled *Transformation and Interpretation in Quantum Mechanics* (p. 83); but he very soon gave them up.

The reason for this mitigated attitude toward the quantum

---

[54] E. Schrödinger, "Discussion of probability relations between separated systems," *Proc. Cambridge Phil. Soc.*, 31, 555–563, 1935.

[55] H. Margenau, "Measurements in quantum mechanics," *Annals of Physics*, 23, 469–485, 1963.

[56] E. Schrödinger, "The meaning of wave mechanics," in: A. George (ed.), *Louis de Broglie physicien et penseur*, op. cit. p. 18.

theory of measurement, made of felt necessity and of incomplete developments, was that Schrödinger had not shaped a very clear idea of what was to be expected from it. On the one hand, he still had the vague hope that one could show how a perturbing operator turns "(. . .) the wave function as time goes on into an eigenfunction of the observable which is measured"[57]; here he showed something like a naive forerunner of decoherence. And on the other hand, he insisted more and more frequently on his idea that for all descriptive purposes, the principle of superposition and the law of (wave-like) evolution is overwhelmingly more important than isolated facts. Thus, Schrödinger sometimes suggested that the solution of the measurement problem was close at hand, provided a proper handling of quantum mechanical descriptions had been achieved. And sometimes he behaved as if he thought that postponing indefinitely the solution of the measurement problem within a perfectly self-consistent quantum theory of measurement was tantamount to ascribing it some (elusive) kind of solution.

Along with this latter trend, namely that of all-pervasive holistic quantum description, some of his sentences are strikingly reminiscent of Everett's interpretation of quantum mechanics. For instance: "Nearly every result (the quantum theorist) pronounces is about the probability of this *or* that or that . . . happening—with usually a great many alternatives. The idea that they be not alternatives but *all* really happen simultaneously seems lunatic to him, just impossible. (. . .) It is strange that he should believe that."[58] But it is also true that Schrödinger's late ideas on the measurement problem could rather have served as an ancestor to many other no-collapse interpretations. His repeated insistence on making a radical separation between the law-like state attribution (which has to be developed indefinitely) and the fact-like value attribution (which must not interrupt the former one) is closer to Van Fraassen's modal interpretation[59] than to Everett's.

To conclude, Schrödinger's late interpretation of quantum mechanics is a remarkably well-designed system, but with a missing keystone. This eagerly sought keystone is nothing more and nothing less than a solution of the measurement problem. However, what Schrödinger suggested in many texts of the 1950s is that the handling of the measurement problem (with its end-product discontinuities) can

---

[57] See this volume, p. 83.

[58] See this volume, p. 19.

[59] B. C. Van Fraassen, *Quantum mechanics, an empiricist view*, Oxford University Press, 1991, p. 275.

be postponed indefinitely without any harm. This way of pushing the measurement problem outside the field of investigation of physics, insisting that its being ignored does not make any difference, clearly does not solve the problem. But, in a Wittgensteinian spirit, it tends to dissolve it.

5 • Sources

Archive for the History of Quantum Physics (AHQP). See T. S. Kuhn, J. L. Heilbron, P. Forman, and L. Allen, *Sources for the History of Quantum Physics*, The American Philosophical Society, Independence Square, Philadelphia, 1967.

July Colloquium 1952: AHQP, Microfilm 44, section 5.

Transformation and interpretation in quantum mechanics: AHQP, Microfilm 43, section 9.

Science, Philosophy and the Sensates (manuscript draft for three "William James lectures" intended to be given at Harvard University, c. 1954): AHQP, Microfilm 44, section 6.

Erwin Schrödinger's personal archive, Alpbach, Austria. (Here I wish to thank Mrs. Ruth Braunizer and her family for their permanent hospitality, help, and friendship.)

Notes for Dublin seminars, 1949, 1955.

Extracts from a letter to A. Eddington, 1940

(Mrs. Braunizer also provided me with excellent copies of the texts reproduced in the AHQP, from the originals in her posession.)

• Typographic Conventions

Schrödinger's handwritten additions to the typescripts are enclosed in braces { }. Erased words, sentences, or paragraphs are enclosed in brackets [ ]. The editor's additions (especially some subtitles) are enclosed in parentheses ( ).

1 • JULY 1952 COLLOQUIUM

1 • Introduction

Let me say at the outset, that in this discourse, I am opposing not a few special statements of quantum mechanics held today, I am opposing as it were the whole of it, I am opposing its basic views that have been shaped 25 years ago, when Max Born put forward his probability interpretation, which was accepted by almost everybody. It has been worked out in great detail to form a scheme of admirable logical consistency that has been inculcated ever since to every young student of theoretical physics.

The view I am opposing is so widely accepted, without ever being questioned, that I would have some difficulties in making you believe that I really, really consider it inadequate and wish to abandon it. It is, as I said, the probability view of quantum mechanics. You know how it pervades the whole system. It is always implied in everything a quantum theorist tells you. Nearly every result he pronounces is about the probability of this *or* that or that . . . happening—with usually a great many alternatives. The idea that they be not alternatives but *all* really happen simultaneously seems lunatic to him, just *impossible*. He thinks that if the laws of nature took *this* form for, let me say, a quarter of an hour, we should find our surroundings rapidly turning into a quagmire, or sort of a featureless jelly or plasma, all contours becoming blurred, we ourselves probably becoming jelly fish. It is strange that he should believe this. For I understand he grants that unobserved nature does behave this way—namely according to the wave equation. The aforesaid *alternatives* come into play only when we make an observation—which need, of course, not be a scientific observation. Still it would seem that, according to the quantum theorist, nature is prevented from rapid jellification only by our perceiving or observing it. And I wonder that he is not afraid, when he puts a ten-pound-note (his wrist-watch) into his drawer in the evening, he might find it dissolved in the morning, because he has not kept watching it.

The compulsion to replace the *simultaneous* happenings, as indicated directly by the theory, by *alternatives*, of which the theory is supposed to indicate the respective *probabilities*, arises from the conviction that what we really observe are particles—that actual events always concern particles, not waves. Once we have decided for this, we have no choice. But it is a strange decision. Observation presents us with two kinds of structural linkage between events which we may distinguish as the *longitudinal* linkage, usually called the particle aspect, and the *transversal* linkage, usually called the wave aspect. Everybody agrees that we encounter great difficulty in trying to reconcile them. This is no doubt so. But if you think of it, you realize the following. In a wave phenomenon you have—not always, but in many cases—the two complementary features of wave- (or phase-) surfaces and of wave-normals of rays. (By complementary I mean nothing mystical, only that they depend on each other; that is: when and if, and to the same degree of approximation as, one of them exists, the other exists). These two features, wave-surfaces and wave-normals, seem capable of representing the two observed features, namely the transversal structural interlacing and the longitudinal one, respectively. A particle phenomenon has not these two traits, only one: the paths or orbits or trajectories. They are capable, at most, of representing the longitudinal linkage, there is nothing in particles to correspond to the transversal one. One might for a moment think of the *forces* between the particles. Everybody agrees that this is not to the point; everybody agrees that there are no forces between photons, moreover that the diffraction and interference of light is the business of every single photon, not of their interaction.

The longitudinal and transversal linkages are of different kind. The first is in a time-like direction and is thus of the ordinary causal type. The other one, the transversal, is a relation between simultaneous world points or at any rate such at a space-like interval, so that there can be no causal relationship between them. It is a relationship of common structure pointing to a common origin. Bertrand Russell, in his treatise on Human Knowledge has enlarged on this kind of structural relationship, which he claims to be at least as important a source of *inductive inference* as the causal relation of permanent succession. A typical example is afforded by all the copies of the same issue of a book or periodical. None of them is either the cause or the effect of any other of them, yet their similarity in structure has a definite and well understood meaning to us. For the points on a wave surface the structural similarity is, that the phase is the same.

Yet the longitudinal and the transversal linkage—*Wirkungszusammenhang* I call it in German—are not sharply delimited, nay they are ever sharply delimited against one another, because, as everybody knows, the wave-surfaces and the wave normals (the rays) are *never* sharply defined. It is true that in some cases one can stipulate an artificial sharp definition; e.g. a complex scalar wave function $\psi(x,t)$ can be uniquely written:

$$\psi(x,t) = A(x,t)e^{i\varphi(x,t)}$$

with A and φ real, and you may call the surfaces $\varphi = $ const. the wave surfaces. But this has a good meaning only when A varies (in space and time) slowly compared with φ.

Now I believe everybody agrees that the *path* or world-line of a particle can be given no *other* meaning than that of a *ray* or (orthogonal) trajectory of the family of wave surfaces. Since these rays are never sharply defined, the paths are at any rate, to say the least, never sharply defined. To me this alone suffices, to strongly question the adequacy of the concept of particle. For what is a particle that has no well-defined path? But we shall later adduce other reasons in the course of this discourse.

For the moment let me still remind you that there are cases when the notions of wave-surfaces and wave-normals break down entirely, and that the chief interest of wave-mechanics is concentrated on these cases. I mean of course the de Broglie wave phenomenon that surrounds the nucleus. Here no trace is left of anything that might be thought of as representing the path of a particle. Hence the idea of point-electrons—whatever it may mean elsewhere—becomes absolutely inadequate in this region, that is to say within the "body" of an atom. To my mind it is patently absurd to call *anything* the probability of *finding* an electron near a particular point in this region—which, by the way, could only mean at a place specified with respect to the nucleus. Nobody has ever tried to *look for one*, nobody ever will; in fact nobody has ever experienced or will ever experiment in this fashion on a single atom of hydrogen or whatnot. What astonishes me most is, that this kind of consideration is adopted as the basis of their theory by a school of thinkers who solemnly proclaim, that they are going to speak of nothing that is not *observable,* and who look with anything between pity and indignation upon him who does.

2 • How It Came About

Allow me now briefly to sketch, how the present situation came about, though I know that I am telling you things well known to all of you—and I am making some use of a type-script written up not just for physicists only.[1]

Max Planck's essential step in 1900, amounted, as we say now, to laying the foundation of quantum theory; it was his discovery, by abstract thought, of a *discontinuity* where it was least expected, namely in the exchange of energy between an elementary material system (atom or molecule) and the radiation of light and heat. He was at first very reluctant to draw the much more incisive conclusion that each atom or molecule had only to choose between a *discrete* set of "states"; that it could normally only harbour certain discrete amounts of energy, sharply defined and characteristic of its nature; that it would normally find itself on one of these "energy levels" (as the modern expression runs)—except when it changes over more or less abruptly from one to another, radiating its surplus energy to the surrounding, or absorbing the required amount from there, as the case may be. Planck was even more hesitant to adopt the view that radiation itself be divided up in portions or light-quanta or "photons", to use the present terminology. In all this, his hesitance had good reasons. Yet only a few years later (1905) Einstein advanced the hypothesis of light-quanta, clinching it with very plausible[2] arguments; and in 1913 Niels Bohr, by taking the discrete states of the atoms seriously and extending Planck's assump-

---

[1] *Editor's note:* This typescript begins with a truncated paragraph which was erased by Schrödinger:

"(. . .) observed facts and have yet lost all interest except to historians. I am thinking of the theory of epicycles. I confess to the heretical view that their modern counterpart in physical theory are the quantum jumps. Or rather these correspond to the *circles* which the sun, the moon and the stars were thought to describe around the Earth in 24 hours after earlier and better knowledge had been condemned. I am reminded of *epicycles* of various orders when I am told of the hierarchy of *virtual* quantum transitions. But let these rude remarks not deter you. We shall now come to grips with the subject proper."

[2] *Editor's note:* the original expression was "irresistible;" it was then erased and replaced by "very plausible."

tions in two directions, [with great ingenuity, but irrefutable consistency]³ could explain quantitatively some of the atomic line spectra, which all are patently *discrete* and has in their entirety formed a great conundrum up to then: Bohr's theory turned them into the ultimate and irrevocable direct evidence, that the discrete states are a genuine and real fact. Bohr's theory held the ground for about a dozen of years, scoring a grand series of so marvellous and genuine successes, that we may well claim excuse for having shut our eyes to its one great deficiency : while describing minutely the so-called "stationary" states which the atom had normally, i.e. in the comparatively uninteresting periods when *nothing happens*, the theory was silent about the periods of transition or "quantum jumps" (as one then began to call them). Since intermediary states had to remain disallowed, one could not but regard the transition as instantaneous; but on the other hand the radiating of a coherent wave train of 3 or 4 feet length, as it can be observed in an interferometer, would use up just about the average interval between two transitions, leaving the atom no time to "be" in those stationary states, the only ones of which the theory gave a description (except perhaps in the lowest one.)⁴

This difficulty was overcome by [quantum mechanics, more especially by]⁵ wave mechanics, which furnished a new description of the *states*; this was precisely what was still missing in the earliest version of the new theory which had preceded wave mechanics by about one year. The previously admitted discontinuity was not abandoned, but it shifted from the *states* to something else, which is most easily grasped by the simile of a vibrating string or drumhead or metal plate, or of a bell that is tolling. [If such a body is struck, it is set vibrating, that is to say it is slightly deformed and then runs in rapid succession through a continuous series of slight deformations again and again. There is, of course, an infinite variety of ways of striking a given body, say a bell,—by a hard or soft, sharp or blunt instrument, at different points or at several points at a time. This produces an infinite variety of initial deformations and accordingly a truly infinite variety of shapes of the ensuing vibration: the rapid "succession of cinema pictures", so we might call it, which describe the vibration following on a particular

---

³ *Editor's note:* this expression was erased by Schrödinger.

⁴ *Editor's note:* hand-written expression.

⁵ *Editor's note:* The bracketed expression was erased lately. This slight, but decisive, modification, can be considered as a symptom of the rising confidence Schrödinger felt in 1952 towards his own interpretation of quantum mechanics.

initial deformation is infinitely manifold. But in every case, however complicated the actual motion is, it can be mathematically analysed as being the *superposition* of a discrete series of comparatively simple "proper vibrations", each of which goes with a quite definite frequency.][6] There is a discrete series of frequencies which depends on the shape and on the material of the body, its density and elastic properties. It can be computed from the theory of elasticity, from which the existence and the discreteness of proper modes and proper frequencies and the fact that any possible vibration of that body can be analysed into a superposition of them is very easily deduced quite in general, i.e. for an elastic body of any shape whatsoever.

The achievement of wave mechanics was, that it found a general model picture in which the "stationary" states of Bohr's theory take the rôle of proper vibrations, and their discrete "energy levels" the rôle of the proper frequencies of these proper vibrations; and all this follows from the new theory, once it is accepted, as simply and neatly as in the theory of elastic bodies, which we mentioned as a *simile*. Moreover the radiated frequencies, observed in the line spectra, are in the new model, equal to the *differences* of the proper-frequencies; and this is easily understood, when two of them are activated simultaneously, on simple assumptions about the nature of the vibrating "something".

3 • The Alleged Energy Balance—A Resonance Phenomenon

But to me the following point has always seemed the most relevant, and it is the one I wish to stress here, because it has been almost obliterated,—if words mean something, and if certain words now in general use are taken to mean what they say. The principle of superposition not only bridges the gaps between the "stationary" states, and allows, nay compels us, to admit intermediate states without removing the discreteness of the "energy levels" (because they have become proper frequencies); but it completely *does away with the prerogative* of the stationary states. The epithet stationary has become obsolete. Nobody who would get acquainted with wave mechanics without knowing its predecessor (the Planck-Einstein-Bohr theory) would be inclined to think that a wave-mechanical system has

---

[6] *Editor's note:* This bracketed paragraph was erased and replaced by the following short sentence: "I need not enlarge on this to you here."

a predilection for being affected by just only one of its proper modes at a time. Yet this is implied by the continued use of the words "energy levels", "transitions", "transition probabilities".

The perseverance in this way of thinking is understandable, because the great and genuine successes of the idea of energy parcels has made it an ingrained habit to regard the product of Planck's constant h and a frequency as a bundle of energy, lost by one system and gained by another. How else should one understand the exact dovetailing in the great "double-entry", book-keeping in nature? I maintain that it can in all cases be understood as a resonance phenomenon. One ought at least to try, and look upon atomic frequencies just as frequencies and drop the idea of energy-parcels. I submit that the word energy is at present used in two entirely different meanings, macroscopic and microscopic. Macroscopic energy is a "quantity-concept" (Quantitätsgrosse). Microscopic energy (meaning hν) is a "quality-concept" or "intensity-concept" (Intensitätsgrösse); it is quite proper to speak of high-grade and low-grade energy according to the value of the frequency ν. True, the macroscopic energy is, strangely enough, obtained by a certain weighted summation over the frequencies, and in this relation the constant h is operative. But this does not necessarily entail that in every single case of microscopic interaction a whole portion hν of *macroscopic* energy is exchanged. I believe one is allowed to regard microscopic interaction as a continuous phenomenon without losing either the precious results of Planck and Einstein on the equilibrium of (macroscopic) energy between radiation and matter or any understanding of phenomena that the parcel-theory affords. If time permits, I shall deal with the question of thermodynamic equilibria later.

The one thing which one has to accept and which is the inalienable consequence of the wave-equation as it is used in every problem, under the most various forms, is this: that the interaction between two microscopic physical systems is controlled by a peculiar law of resonance. This law requires that the *difference* of two proper frequencies of the one system be equal to the difference of two proper frequencies of the other;

$$\nu_1 - \nu_1' = \nu_2' - \nu_2. \qquad (1)$$

The interaction is appropriately described as a gradual change of the amplitudes of the four proper vibrations in question. People have kept to the habit of multiplying this equation by h and saying it means, that the first system (index 1) has dropped from the energy level hν_1 to the energy level $h\nu'_1$, the balance being transferred to the second system, enabling it to rise from $h\nu_2$ to $h\nu'_2$. I deem this interpretation to be obsolete. There is nothing to recommend it, and it bars the understanding of what is actually going on. It obstinately refuses to take stock of the principle of superposition, which enables us to envisage simultaneous gradual changes of any and all amplitudes without surrendering the essential discontinuity, where it prevails, namely that of frequencies. Of course the condition of resonance may include three or more interacting systems. It may e.g. read

$$\nu_1 - \nu'_1 = \nu'_2 - \nu_2 + \nu'_3 - \nu_3. \qquad (2)$$

Moreover we may adopt the view that the two or more interacting systems are regarded as *one* system. One is then inclined to write equations (1) and (2), respectively, as follows

$$\nu_1 + \nu_2 = \nu'_1 + \nu'_2 \qquad (1')$$

$$\nu_1 + \nu_2 + \nu_3 = \nu'_1 + \nu'_2 + \nu'_3 \qquad (2')$$

and to state the resonance condition thus: the interaction is restricted to constituent vibrations of the *same* frequency. This is a familiar state of affairs, of old. Unfamiliar is the tacit admission that frequencies are additive, when two or more systems are considered as forming one system. It is an inevitable consequence of wave mechanics. Is it so very repugnant to common sense? If I smoke 25 cigarettes per day, and my wife smokes 10, and my daughter 12,—is not the family consumption 47 per day—on the average? {Anyhow, it shows that wave mechanics is far from reducing everything to a "classical" picture}.

4 • Removing Some Difficulties

But [jokes aside], why has this apparently simple and obvious view been universally rejected? I propose it now afresh—though with a relevant modification: so-called second quantization or field-quantization has to be worked into it. Let me take this for granted without going into details, it would take us too long. But I must enter on the momentous grounds for rejecting it, I must try to dispose of them.

The view has never been taken seriously, after the highest authorities at Copenhagen and Göttingen had pronounced against it. So it came about that its most frightening consequence has, to my knowledge, never been pointed out. I wish to point it out at once, but discuss it later. According to Einstein mass and energy are the same thing.

Hence there can be no shadow of a doubt, that the elementary particles themselves are Planckian "energy parcels". This is fine. But if we now dismiss the idea—as too naïve—the idea, that energy is always exchanged in whole parcels, if we replace it by resonance view, does this not mean that atomism will go by the board? Well no, not atomism, only the corpuscles, the atoms and the molecules, but not atomism. I believe the discrete scheme of proper frequencies of second quantization to be powerful enough to embrace all the actually observed discontinuities in nature for which atomism stood, without our having to enhance them by fictitious discontinuities that are not observed.

But let us first consider simpler questions: some typical experiments which allegedly enforce the energy parcel view, while to my mind they do not. A beam of cathode rays of uniform velocity, which can be gradually increased, is passed through sodium vapour. Behind the vessel containing the vapour the beam passes an electric field, which deflects it and tells us the velocity of the particles after the passage. At the same time a spectrometer inspects the light, if any, emitted by the vapour. For small initial velocity nothing happens: no light, no change of velocity in the cathode beam. But when the initial velocity is increased beyond a sharply defined limit, two things happen. The vapour begins to glow, radiating the frequency of the first line of the "principal series"; and the beam of cathode rays emerging from the vapour is split into two by the deflecting electric field, one indicating the initial velocity unchanged, and another slow one, that has "lost an amount of energy" equal to the frequency of the said spectral line multiplied by Planck's constant h. If the velocity is further increased the story repeats itself when the incident cathode ray energy increases beyond the "energy level" that is responsible for the second line (or rather the "level-difference" in question); this line appears and a third beam of cathode rays with correspondingly reduced speed occurs; and so on. This was and still is regarded as blatant evidence of the energy parcel view.

But it is just as easily understood from the resonance point of view. A cathode ray of particles with uniform velocity is a monochromatic beam of de Broglie waves. Only when its frequency (v_1) surpasses the frequency difference $(v'_2 - v_2)$ between the lowest (v_2) and the second (v'_2) proper frequencies of the sodium atom, is there a de Broglie frequency $v'_1 > 0$ that fulfils the resonance demand, equation (1). Then the vibration v'_1 appears in the de Broglie wave and v'_2 among the atoms which begin to glow with frequency $v'_2 - v_2$, since Maxwell's "electromagnetic vacuum" is prepared for resonance with anything. The splitting of the cathode ray beam in the deviating electric field, after passing the vapour, is accounted for by de Broglie's wave equation. An electric field has for de Broglie waves an "index of refraction" that *depends* on their frequency ["dispersion"] and has a gradient in the direction of the field (which thus acts as an "inhomogeneous medium"). Any further events that might happen, for instance a transfer of some of the "energy quanta" $h(v'_2 - v_2)$ from the sodium atoms to other gas molecules by "impacts of the second kind", are just as easily understood as resonance phenomena, provided only one keeps to the wave picture throughout and for all particles involved.

Many similar cases of apparent transfer of energy-parcels can be reduced to resonance—for instance photochemical action. The pattern is always the same: you may either take equations like (1) or (2) (between frequencies) as they stand (resonance), or multiply them by h and think they express an energy balance of every single microtransition. In the preceding example one point is of particular interest. One is able by an external agent (the electric field) to *separate in space* the two or more frequencies which have arisen in the cathode ray by the interaction; for they behave differently towards this agent and the different behaviour is completely understood from de Broglie's wave equation; one thus obtains two or more beams of homogeneous frequency (or velocity). It is extremely valuable that there are simple cases of this kind in which the separation into two "phases" has nothing enigmatic; it is an immediate consequence of the principles laid down in L. de Broglie's earliest work on matter waves. I say, this is fortunate; for there is a vast domain of phenomena in which the separation in space either takes place in the natural conditions of observation, or can easily be brought about by simple appliances; but it is not as easily explained on first principles. This might dishearten one in accepting the view of gradually changing amplitudes, that I put forward here; for the separation into different phases that produces itself before our eyes seems to confirm the belief, that a discontinuous abrupt and *complete* transition occurs in every single microscopic interaction.

5 • Chemistry, Photochemistry, and the Photoelectric Effect

The vast group of phenomena I am alluding to is in the first place ordinary chemistry. Two or more constituents, mixed in a solution or in a gaseous phase, begin to react with each other, under the influence of light or otherwise; the portions that have reacted and have formed a new chemical compound may separate themselves almost entirely from the rest and form a new phase, say because the product is almost insoluble in the liquid, or (in the case of a gaseous mixture) by its being a liquid or solid with a low vapour pressure at the temperature in question. Almost any chemical reaction may serve as an example, but let us take a slow one to facilitate speech and thought. If a suitable mixture of hydrogen gas (H_2) and oxygen gas (O_2) is illuminated by ultraviolet light, the following slow reaction is induced

$$2\,H_2 + O_2 \to 2\,H_2O. \tag{3}$$

As the concentration of water vapour (H_2O) increases, part of it separates off into liquid droplets.

The actual process is not as simple as the balance (3) indicates, it is a chain reaction. But we need pay no attention to this, and contemplate only the initial state and the end-product. Wave-mechanically the gaseous mixture is represented by a vibration of the combined system, and, by the way, not by *one* proper vibration since there is anyhow the vast variety of translational and rotational modes, and, of course, the electronic modes. The gaseous compound, H_2O, is represented by an entirely different vibration of the *same* system. The modes composing it, absent at first, are gradually chiming in as the reaction proceeds. But then there is a *third* group of vibrations, representing the liquid H_2O; they gradually build up where they are facilitated by dust nuclei, and observed as droplets. It is, of course, regrettable that wave mechanics does not allow us to follow this *observed* process analytically, while, in the now current interpretation, ample information is forthcoming about a host of experiments that nobody has ever been or ever will be able to perform (for instance we are told, what is the probability of our finding at a definite spot inside a given hydrogen atom and electron, if we look for one). But there is no reason to suspect that the separation of phases is fundamentally different from the spectroscopic resolution of a beam of light or of cathode rays into its monochromatic constituents. One need not be afraid that the formation of spatial boundaries, separating coherent regions of chemically or physically distinct properties, cannot possibly be controlled by the wave equation, but must necessarily be accounted for by the picturesque pageantry of individual molecules swallowing or re-spewing whole energy parcels, being disrupted and re-combined, until they eventually go to form one or two molecules of a new type.

I deem the latter simply wrong; it is not in accordance with our present state of knowledge, whose further progress is hampered if these easy pictures, that are in common use, are taken literally. And we are encouraged to take them literally not only by text-books and popular essays but also by the language used in very high-browed technical treatises. By this I will not deny that this imagery is a very useful, nay indispensable, conceptual shorthand in chemical research. One cannot see how to avoid it, when e.g. a complicated chain reaction is to be unravelled. And, of course, the chemical equation for describing a reaction will never be ousted, though it ostensibly describes the single micro-event and is wrong in this. It is an instance of the famous "as if". It is not the first instance of this kind in the relation of chemistry and physics. The chemist used the valency stroke for building models of complicated molecules. It represented very real facts of observation. For a long time the physicist could not afford any explanation of the mechanism of the chemical bond. Then, in brief succession, *two* were given: there is a heteropolar bond (Kossel 1916) and a homopolar bond (London-Heitler 1926). The discoveries were illuminating to the chemist, indeed they removed some difficulties caused by interjecting the valency strokes too naïvely. But, of course, the valency strokes were retained as an extremely convenient shorthand. They could be retained because they were based on carefully pondered observation.

As one of the simplest photochemical reactions we may regard the photoelectric effect, which was one of the main incentives for Einstein in 1905 to launch the hypothesis of light quanta. When a metal plate is illuminated by light of sufficiently high *frequency*, electrons emerge from it forthwith with an energy corresponding to this frequency. There is no time delay, even when the *intensity* of the incident light is so weak that according to the electron theory of H.A. Lorentz, which was at the time in full swing, an electron would need half an hour to be sped up to the velocity in question. This was—and, I am afraid, still is—regarded as convincing evidence of the instantaneous transfer of whole quanta of energy from the light to the electron. You all know the present orthodox interpretation to be as follows. The incident light beam produces at once in each of tens of thousands of electrons an exceedingly small *probability* of taking within the next split second a leap into a state of higher translational energy; a correspondingly small fraction of those tens of thousands do so and emerge from the metal, and that is why the game starts without delay.

But according to wave mechanics, as put forward by de Broglie and myself and generally accepted, the interaction does produce without delay electronic wave trains of the higher frequency that we observe emerging from the metal [(For to observe the *frequency* of an electron or its *velocity* means the same thing)]. After this has been recognized, is the probability scheme any longer needed? Has the idea of the mysterious sudden leaps of single electrons not become gratuitous? Is it expedient? The waves are there anyhow, and we are not at a loss to prove it. We need only put a tube of crystal powder in the way of the emerging beam and produce an interference pattern of the type first achieved by G.P. Thomson (it might be not as beautiful as Thomson's, but it would vouch for the waves all the same).

6 • (Individuality and Sameness)

I said earlier that I do not think atomism will go by the board when we give up believing in quantum jumps or "exchange of energy parcel-wise". Remember the argument that seems to endanger atomism: the elementary particles themselves are energy parcels mc^2. So if the energy parcels are not to be taken seriously, neither are the elementary masses.

Here we must make a careful distinction. If I am told the mass of a

$$\left.\begin{array}{l}\text{Proton is } 1.00758 \\ \text{Neutron is } 1.00898\end{array}\right\} \times 1.6603 \cdot 10^{-24}\,\text{g},$$

I take it as seriously as when I am told the difference between the 3rd and the 2nd energy level of the hydrogen atom is so and so many e-volt, or equivalently the wave-length of H_α is so and so many Ångström.

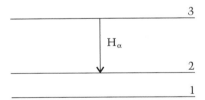

I take the statement about H_α and that about m_p as informations of exactly the same kind. What I do not take seriously is the existence of single small individuals or mass m_p; just as little as I (or for that matter anybody else) imagines that a hydrogen atom contains a little body of mass:

$$R_y(1/2^2 - 1/3^2) \cdot h/c. \quad (R_y = \text{Rydberg constant in cm}^{-1}).$$

There is a general reason for not regarding the elementary particles—electrons, protons, light-quanta, mesons—as individuals, a reason very well known to everybody. When you are dealing with a system that contains equal particles you must wipe out their individuality, lest you get quite wrong results. This is so whether the number is small or large. It is so for the two or three electrons of the He or Li atoms, respectively. You must use only wave-functions that are antisymmetric with respect to the coordinates of the 2 or 3 electrons—or you must use the Fermi-Dirac type of field quantization, which amounts to the same. And if you are dealing with a palpable body of a gas containing, say, 10^{20} molecules, you must use functions symmetric or antisymmetric with respect to all the 10^{20} corpuscles—or use the appropriate quantization of their field, which amounts to the same.

This means *much more* than that the particles or corpuscles are all *alike*. It means that you must not even *imagine* any one of them to be *marked*—"by a red spot" so what you could recognize it later as *the same*. If you happen to get 1000 of more records of a proton, as you often do, then notwithstanding the greatest psychological urge to say: it is the *same* proton, you must remain aware, that there is no absolute meaning in this statement. There is a *continuous* transition between cases where the sameness obtrudes itself to such where it is *obviously* meaningless.

As against this, *waves* can easily be marked, by their shape or modulation. If you hear a good friend speaking on the wireless at New-York, you can tell with dead certainty that the wave that hits your receiver is the same which his voice has modulated many 1000 miles away. The light-waves that tell you something about what happens in a protuberance on the sun's limb are the same that were emitted there. If your wife shouts up to you from the garden, you recognize her voice and you know that the sound waves which hit your ear are the same that she has produced by her vocal chords. These are trivial macroscopic examples. But the waves of quantum mechanics exhibit the same feature. They *have* to be treated as individuals. You know that if you deal with such many-body problems as the He-molecule or a gas of 10^{20} molecules, by the method of wave-mechanics, the proper modes have to be regarded as distinguished from one another, they have to be treated as true individuals. You must not apply Fermi-Dirac statistics or Bose-Einstein statistics to *them*, but ordinary Boltzmann statistics: *then* you obtain the correct results, the same as you get by applying the new-fangled statistics to the non-individual corpuscles.

There are other instances for abrogating individuality to the particles. I will mention the well-known and often quoted interference-experiment with the two slits:

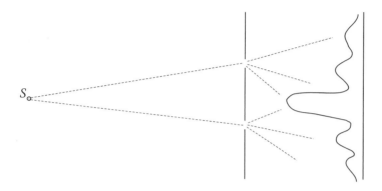

The interference pattern is composed of single-particle events, scintillations or reduced silver-grains. You cannot trace the path of anyone of these alleged particles back to the source. You cannot tell through which slit it was launched. If you could tell it for one, you could in principle tell it for everyone. But then you cannot understand the pattern that is produced.

Another point. Consider a system composed of N equal particles of mass m each. The *bulk* of its energy is

$$Nmc^2$$

If you dismiss the view that it is always in a sharp energy state (as I propose it to be dismissed) then the number N is also not sharp. I suppose that it usually has a spread of order $\sqrt{N}$. If this is accepted, it does away with something that I have always regarded as a serious difficulty, though Heisenberg dismissed it lightly in 1928. You determine the place of a particle at one moment with great precision (A), making its momentum p thoroughly blurred as to absolute value and direction. What follows this must be pictured as a spherical wave issuing there and then and spreading in all directions. (You see I am following for the moment the current view).

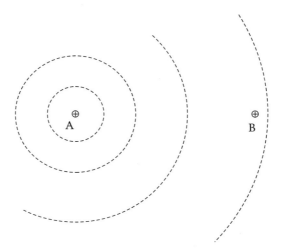

Now if you have distributed a great number of detectors at some large distance, you may find your particle at B. Then the following difficulties arise:

(i) The current view assumes that by this the spherical wave collapses into a small wave parcel near B. This is strange enough. However to the current view the wave is only a probability wave, a wave of information; that is how the current view disposes of the objection that the collapsing involve a propagation of action with more than the velocity of light. But there are further difficulties:

(ii) You can very easily determine the place B with an accuracy such that the quotient AB/τ (τ: the time between the two observations) affords a determination of momentum far too precise when compared with the accuracy of location *at A*. It is hard to admit that you have only produced this relatively sharp momentum at B since it is in accordance with *time* and with the *direction* the particle has taken from A to B. Moreover,

(iii) You can, of course, make the location at B so sharp, that the value AB/τ also contradicts the precision of the location *at B*. (Of course this will *not* be the momentum *after* the measurement B.) And finally:

(iv) You can carefully balance the {simultaneous} accuracies of a simultaneous momentum- and a place-measurement at B, in such a way that the *two* momentum determinations

$$AB/\tau\,;\quad p_B$$

agree fairly well and are "much too precise", when held against the location at A.

What produces these antinomies? It is well known that the Heisenberg uncertainty principle is, in wave-mechanics a truism. *Wave mechanics can never conflict with the U.P.* The conflict is produced by imposing on wave mechanics the probability view. This view entails the ruling that *on measurement* the validity of the wave equation is momentarily suspended and the wave function changes abruptly in a way not controlled by, nay strictly at variance with, the wave equation, in our case: the collapsing of the whole spherical wave into a small wave-parcel. That is how we obtain—in two different ways—an accuracy of momentum which conflicts with the accuracy of location obtained at A: by violating the principles of wave mechanics. To my mind it is patently absurd to let the wave function be controlled in two entirely different ways, at times by the wave equation, but occasionally by direct interference of the observer, *not* controlled by the wave equation.

If, in the present case, you wish to avoid the paradox while keeping as closely as possible to the "particle-language" you need only accept that the number N of particles is not sharp but has a spread of order $\sqrt{N}$. This means that the wave contains proper modes not only for $N = 1$, but also for $N = 2, 3, \ldots$, the lower ones being still fairly strong. Then, according to recognized rules, the "finding of a particle at B" leaves the conditions for finding one somewhere else *unchanged*: the wave does *not* collapse, the paradox is avoided.

You may object that in some cases, after all, we *are* experimenting with single particles. You may point to the linear traces or trajectories in the photographic plate and in the Wilson chamber, the collisions, the stars, the decaying particles etc. This range of phenomena, very fashionable now, seem to form an unconquerable stronghold for the particle view. I shall summarize my attitude towards these things in a few brief remarks.

(i) We do not in these cases *experiment* with single particles; we are scrutinizing records. We cannot repeat any of them under planfully[7] varied conditions—the typical procedure of the experimenter.

(ii) These records, in their apparent simplicity, appeal to the vivid imagination of an intelligent child. The great warriors for atomism, from Democritus down to Dalton and Boltzmann would have gone into raptures about them. But they are certainly not as simple as they look at first sight. This is witnessed by the pages and pages of intricate formalism that the current theory devotes to accounting for even the simplest of them, {e.g. the Klein-Nishima formula of the Compton effect}.

(iii) I think it is fair to call them an exceptional group of events. They need very special contraptions (however familiar they have become to us by now) and they only occur with very fast or high energy particles.

(iv) They are typical for what I called the *longitudinal* correlation of linkage. To the naïve onlooker they strongly suggest the particle view. But we ought not to let this prevail over our minds. The actual theoretical computations use waves. If we turn our eyes away from this special group of facts to the whole of physics we find the two kinds of linkage, transversal and longitudinal. Both have a counterpart in the wave aspect—in the wave surfaces and rays—but the particle aspect has only the one (longitudinal).

Any *modulation* that occurs in a wave front marks it and spreads in the direction of the ray. This may be the nature of these, after all, rather ephemeral particle tracks.

7 • Thermodynamics

I want to sketch briefly how thermodynamics is deduced from statistics without the "transfer of energy parcelwise". The identification of statistical concepts with thermodynamical concepts *always* requires some basic assumptions. We need three.

(i) With any normal mode ("eigenfunction of the Hamiltonian") of a big system, that is one with many degrees of freedom, we associate the energy h times its frequency, (the latter being

---

[7] *Editor's note:* "planfully," as used by Schrödinger in this sentence, presumably means "according to a plan."

in many cases additively composed of the frequencies of its constituents as wave mechanics will have it,), say E_r.

(ii) We also associate with it an entropy

$$S = k \log m_r$$

where m_r is the degeneracy of the normal mode in question.

(iii) In the long run and on the average the absolute squares of the amplitudes ("excitation strength") of the m_r normal modes belonging to the same E_r are equal.

The last assumption is as open to aggression as any assumption of statistical disorder is. (Invariant!)

Now envisage two systems

Eigenvalue	multiplicity
E_r	m_r
F_s	p_s

The first is *the* system you investigate, the 2nd, the heat bath, shall be enormous. You assume them to be in loose interaction, and at the *total* eigenvalue

$$E = E_r + F_s .$$

To practically every E_r there will be an F_s satisfying this condition. For any particular E_r there will be a number of

$$m_r p_s$$

eigenfunctions. From (iii) this is a relative measure of the excitation strength of E_r. From (ii)

$$k \log p_s = S(F_s) = S(E - E_r)$$
$$= S(E) - \partial S/\partial E \cdot E_r = S(E) - E_r/T$$

Hence

$$m_r p_s = m_r e^{-E_r/kT} \cdot e^{+S(E)/k}$$

This is the canonical distribution of amplitude squares of our system (the last factor is independent of index r).

2 · TRANSFORMATION AND INTERPRETATION IN QUANTUM MECHANICS (C. 1952)

• Introductory Remark

These lectures,[1] though they are concerned with simple elementary aspects of quantum mechanics, are not intended to serve as a first introduction to the subject. Their main objective is to explain the logical connections and the epistemological meaning of the various assumptions that form the foundation of the theory as it is looked upon today. It is assumed, the reader knows that there is such a thing as the elementary wave function, function of the configuration coordinates; that it can be transformed to other variables and has some meaning in every form. He may also know that this wave function changes as time goes on, but is also supposed to change when the system is interfered with physically by a so-called measuring process. He would like to know, so we assume, precisely what meaning is attributed to the wave function in its various forms, and what induces one to adopt exactly this interpretation; also what suggests the particular time-change of the wave function that we admit in a system that is not interfered with, and the changes we have to admit as the result of a physical interference; and what induces one to admit these latter changes, about which we should also like to know, to what extent they mean physical changes in the object or only changes in our knowledge about the physical object.

A general notion of complete orthogonal sets of functions, discrete or continuous sets as the case may be, is presupposed; also that the eigenfunctions of certain linear operators do form such complete

---

[1] *Editor's note:* That this set of lectures was written in (or shortly after) 1952 is made likely by a reference to J. Von Neumann's *Mathematical foundations of quantum mechanics* of 1932 as having been published "about twenty years or more" earlier (p. 69 of the typescript).

orthogonal sets. The detailed investigation of the precise mathematical conditions in which this is the case is outside the scope of these lectures, which though they deal throughout with mathematical entities and their relations are yet principally aimed at clearing up the physical situation.

1 • The Wave Function; Linear Operators; Eigenfunctions and Eigenvalues

The wave function in its naïve form is a function of the configuration coordinates of a physical system. We say: of the values the coordinates can take on. For most general purposes it is sufficient to speak of one coordinate only, and even to have in mind a system whose configuration is exhaustively described by one coordinate x. We shall adopt this view, though the reader is perfectly allowed in all the following to think of x as indicating symbolically a finite number of configuration coordinates. The values which x can take on, ranging as a rule from $-\infty$ to $+\infty$ will also be called its eigenvalues. In calling them so, one regards x also as a linear operator, operating on the wave function, viz. the operation "multiplying by x." It is important to distinguish between the operator x and the ordinary algebraic variable x, representing the entirety of the operator's eigenvalues, on which the wave function depends. It is precisely the difference between a matrix and its eigenvalues, that is its diagonal terms when the matrix is, by a suitable transformation, "made diagonal." It is only when the matrix happens to be on the diagonal—as the operator x is in the present case—that one is prone to confuse the two things. (We shall proceed in due course to explain in a general way the technical terms used in these statements). According to the two meanings of x, a function of x, say f(x), has also two entirely different meanings. A function of the algebraic parameter is precisely the thing we called a wave function. But a function of the operator x is itself an operator on a wave function, viz. multiplying it by that function of the parameter x. Moreover a wave function can have any form, in particular it might, as a special case, be just x. So we have really to distinguish between *three* meanings of x:

x as an operator (representing, as we shall see, an observable),
x the set of eigenvalues of the former.
x as a special case of a wave function on that set.

We proceed to define some of the technical terms used in the preceding. An *operator* is a prescription for associating with every function, defined in a certain domain, another function, which may be, and very often is, in the same domain, but may be in another domain characteristic of the operator in question. A *linear* operator, say ϑ, is one for which the general rule holds that

$$\vartheta\left[C\varphi(x) + C'\psi(x)\right] = C\vartheta\varphi(x) + C'\vartheta\psi(x) \tag{1,1}$$

where φ and ψ are any two wave functions and C and C' any two constants. Particularly simple examples are the *operator* x, which associates with the wave function $\varphi(x)$

$$\vartheta\varphi(x) = x\varphi(x)$$

and the *operator* f(x) which associates with the wave function $\varphi(x)$ the *wave function* f(x)φ(x). But the class of linear operators is a very general one. It is well to realize this from the outset. The following associations of functions (indicated by an arrow →) are examples of linear operations:

$$\varphi(x) \to \frac{d\varphi}{dx}$$

$$\varphi(x) \to \frac{d^n\varphi}{dx^n}$$

$$\varphi(x) \to \varphi(x+h) \qquad (h = \text{a constant})$$

$$\varphi(x) \to \varphi(x^3)$$

$$\varphi(x) \to \varphi(x^2 + 2ax + b) \qquad (a, b = \text{constants})$$

$$\varphi(x) \to \int K(x, x')\varphi(x')\, dx'$$

$$\varphi(x) \to a\varphi(x+h) + b\varphi(x^3) \qquad (a, b, h = \text{constants}).$$

More generally h, a, b in these examples need not to be constants but might be given functions of x. In particular then *any* substitution, or even substitutions, even if φ depended on more than just one variable x, and even when combined linearly by given coefficient functions, belong to our class, as is easily proved by considering the condition (1,1). Envisage for instance the transformation of a tensor field, say A^r_{st}, by an *arbitrary* transformation in the General Theory of Relativity. The components in the new frame, say A'^r_{st} may be said to be associated to A^r_{st} by a linear operator, provided one takes the point of view, that

A_{st}^r is a function not of 4 but of 7 variables, viz. the 4 coordinates and the three indices, r, s, t.

It is a useful exercise to investigate in some of the simpler examples, what functions would be *reproduced* by the operator up to a multiplying constant (this is one of the demands put on an *eigenfunction*). For example, with h and λ some constants, what functions $\varphi(x)$ fulfil

$$\varphi(x+h) = \lambda \varphi(x) ?$$

In this case obviously

$$\lg \varphi(x+h) = \lg \lambda + \lg \varphi(x);$$

that is to say the logarithm of φ has to be an "additively periodic function," as e.g. an elliptic integral, or also a function like

$$x \frac{\lg \lambda}{h} + \cos \frac{2\pi x}{h}.$$

It is useful to note that *linearity* has nothing to do with *simplicity*. It is easy to indicate very simple associations which are *not* linear, e.g.

$$\varphi(x) \to [\varphi(x)]^2$$

$$\varphi(x) \to \varphi(x) \frac{d\varphi}{dx}.$$

We now have to dwell on the notion of eigenfunctions of an operator. One of the demands is that it be a function that is reproduced by the operation up to a multiplying constant. This mutiplying constant is called the eigenvalue of that operator belonging to this eigenfunction. However the most interesting fact about the operators we are principally concerned with in quantum mechanics, viz. that only a selected discrete set of constants turn up as possible eigenvalues, is not brought about by the "reproduction demand" as we put it, but by additional demands as to the *continuity* of the function itself and of its derivatives up to a certain order, in conjunction with the demand of finiteness or at least quadratical integrability. For instance the well known eigenvalue equation

$$\left(-\frac{d^2}{dx^2} + x^2 \right) \varphi(x) = \lambda \varphi(x) \tag{1,2}$$

has, of course, a couple of solutions for every value of the constant λ. But one can see immediately that outside the limited region in which

$x^2 < \lambda$ (taking λ to be real and positive) the φ-curve is convex towards the x-axis and therefore "in grave danger" to go off to $\pm\infty$ as x does. Only when λ is a positive odd integer there is a solution that weathers these dangers, for instance $\varphi = e^{-x^2/2}$ for $\lambda = 1$, and $\varphi = xe^{-x^2/2}$ for $\lambda = 3$. Hence these are called eigenvalues. But their prerogative is founded solely on the demand of continuity of *φ and its first derivative*. Otherwise one could for almost any value of λ pick out a solution that has two zeros, and "use" it only between these two points, replacing it by zero outside. Its only defect would be that it has two "kinks" where the first derivative is discontinuous. The best reason one can give for excluding this is to say, that the condition (1,2) must be satisfied *everywhere* without exception of even a single isolated point: for indeed, where the 1st derivative changes abruptly, the 2nd does not exist or, if anything, is infinite. *Our demand will thus depend on the highest order of derivative, if any, involved in the operator.*

What are the eigenfunctions of the simple operator "multiplying by x" from which our considerations started, and why did we call any value of this continuous variable an eigenvalue of that operator? If $x\varphi(x)$ is to be, identically in x, a constant multiple of $\varphi(x)$, say $x'\varphi(x)$, then we must have, identically in x,

$$(x - x')\varphi(x) = 0 \qquad (1,3)$$

It follows that φ must vanish everywhere except at the point $x = x'$. Hence the function must be discontinuous; but since our operator does not involve even the first derivative, this does not violate the principle reached in the preceding paragraph; and we have not choice. Unhappily we must do worse than that. The customary notation for the square of the eigenfunction in question is

$$\delta(x, x'),$$

and this function is supposed to yield the value *1* when integrated over an x-interval that contains the point $x = x'$ in the interior. We shall not dwell here on the conceptual difficulties this δ-function involves; it cannot be avoided without enormous circumstantiality.

We must still explain a more general aspect of linear operators, without giving detailed proofs. It can be shown that the function ψ which a linear operator (i.e. one that satisfies (1,1)) associates with a given φ is always obtained from φ in the following manner: the value of ψ at any particular point of its domain (call this domain y, because it may be different from the domain of φ, which we shall call x)—I say the value of ψ at any particular point y is obtained as a linear aggregate of, in general, *all* the values of φ, each multiplied by a certain number, which *does not depend on the particular* φ but only on the places y and x of the two domains. The entirety of these numbers characterizes the operator exhaustively. It is called its matrix, an expression that is borrowed from the case when both domains, y and x, are discrete point sets rather than continuous variables (in the latter case one is more inclined to speak of a function of two variables).

If the two domains are identical, it may happen, that the operator forms the associate $\psi(x)$ of *any* $\varphi(x)$ not really from *all* the values of $\varphi(x)$ but only from the value of φ at the same point x. Such an operator is said to be "on the diagonal" (a term that again alludes to the case when x is a discrete domain). Another thing that may happen is, that the "matrix element" determining the contribution of $\varphi(x')$ to $\psi(x'')$ is the same as (or, if complex, is the complex conjugate of) the matrix element, determining the contribution of $\varphi(x'')$ to $\psi(x')$. Such an operator is called *Hermitian*. It has the valuable property that it can always be *made* diagonal (by the kind of transformation we are about to study in the next section) and that its diagonal elements (that is to say its eigenvalues) then always turn out real. A *differential* operator is Hermitian and is, in loose quantum mechanical terminology, called "self-adjoint," when its adjoint operator is its complex conjugate. (To explain "adjoint": two differential operators D and D' are called adjoint when f D'g − g Df has the form of a "divergence," that is consists of a sum of partial derivatives with respect to the independent variables, and that for *any* two functions f and g). —Some of the facts, very briefly collected in the last few paragraphs, will, so I hope, become clearer in the course of the investigations to follow.

2 • The Transformation to Another Frame

We return to the first paragraph of the previous section. The whole aspect given there is a very special one. There is an enormous variety of different "frames" to which it can be transformed. This transformation means in the first place that the wave function becomes a function of another parameter, namely of the eigenvalue of *another linear operator*. This operator is nearly always chosen Hermitian, so that its eigenvalues are real; the wave function remains therefore a function of a real argument, though quite often it is a complex function. The class of transformations we have to consider is much more general than to regard the wave function as a function not of x, but of some given function of x; this is a special case of our transformation, but a very trivial one. It would mean that we take as that *other* operator simply a given function of the *operator* x, say the *operator* f(x). This operator is already on the diagonal in the original x-frame, its eigenvalues being simply the *numbers* f(x). The whole process is nothing but the familiar change of the configuration coordinate, or coordinates, as for instance when we go over from Cartesian coordinates to polar coordinates. Within the class of transformations we have to consider now, this change, which may be called a point-transformation, is so trivial, that a frame of reference obtained from a given frame in this simple fashion may for all intents and purposes be regarded as the *same* frame.

From this it transpires, that in the general transformation it is not the new set of eigen*values* that matters, but the set of eigen*functions* of the new operator. Indeed the former can be changed by the kind of trivial transformation indicated above into almost any set that we choose in one-one correspondence to it, without altering the eigenfunctions. We can see this clearly in the configuration frame where $\delta(x, x')$, the eigenfunction of the *operator* x, is, of course, also an eigenfunction of the *operator* f(x), belonging to the eigenvalue f(x').

In fact, for forming an idea of the vast manifold of transformations we wish to contemplate, or for indicating a particular one of them, there is no need to turn one's mind to the host of eigenvalue problems of all possible new operators and to select one of them for which we happen to be able to solve the problem. We may indicate any such transformation by giving ourselves directly a complete set of orthogonal functions in the original x-frame, a set of functions then of the configuration parameter x and of a number ℓ that labels the functions within the set. This label may be continuous or discrete, but to fix the ideas we shall for the moment take it to be discrete. In this case we may without loss of generality take for ℓ the natural numbers, either 1, 2, 3, 4, ... or 0, 1, 2, 3, 4, ... This label ℓ takes the place of the eigenvalues of that "new operator" of which we spoke in the original setting of the task, while the transformation is indicated by a (possibly complex) *function of two variables*

$$f(x, \ell), \qquad (2,1)$$

to wit of the eigenvalue x in the old frame and the eigenvalue ℓ in the new frame. It is convenient to supplement the assumption of mutual orthogonality by normalizing to 1 the integral over x of the absolute square of each function. This results in

$$\int f^*(x, \ell) f(x, m) \, dx = \delta_{\ell m} \qquad (2,2)$$

the δ meaning the Kronecker symbol. To prove that there is a certain reciprocity with regard to the two arguments of f—the two sets of eigenvalues—one may state, along with (2,2)

$$\sum_\ell f^*(x, \ell) f(y, \ell) = \delta(x, y) . \qquad (2,3)$$

This is indeed true for any orthonormal set provided one connives at the fact that the sum diverges—which it cannot help to do if it is to represent the queer δ-function. Anyhow when multiplied by any $\varphi(x)$ and integrated over x it formally gives $\varphi(y)$, which is all that the δ-function has to do.

Generally speaking the *analytical form* of these two conditions of orthonormality—I mean, whether there is an integral or a sum on the left and a Kronecker-δ or a Dirac-δ on the right—depends on the continuity or discreteness of the two sets of eigenvalues, x and ℓ, respectively; the reader will easily make out for himself. Some authors have introduced special notations to embrace all possibilities. I am averse to this simplification which rather tends to let us disregard mathematical difficulties that may turn up in special cases. For general considerations we shall, to fix the ideas, continue to regard x as continuous and ℓ as discrete.

3 • The Transformation of the Wave Function

A wave function in the new frame is any function of the label ℓ, thus (if ℓ is discrete) a set of, possibly complex, numbers a_ℓ. We have not yet indicated the correspondence. The guiding principle is that we want linear relations with constant coefficients to be conserved, and the inner product of $\varphi(x)$ and $\psi(x)$, which is defined by

$$(\varphi, \psi) = \int \varphi^*(x) \psi(x) \, dx \qquad (3,1)$$

to be invariant. (This makes orthogonality and normalization invariant properties!). This is achieved by defining the set of numbers a_ℓ

$$a_\ell = \int f^*(x, \ell) \varphi(x) \, dx \qquad (3,2)$$

as the function φ in the new frame. Notice

$$\varphi(x) \equiv \sum_\ell a_\ell f(x, \ell). \tag{3,3}$$

There is complete analogy, for function-space, with rotations in ordinary space, or rather with their complex generalization as unitary transformations. The $f(x, \ell)$ are the direction-cosines between the old and new axes. (It is good to observe, that "unitary" is not the *only* complex generalization of "real orthogonal." The one that results in 4-dimensional Minkowski space by putting $x_4 = ict$ is *not* unitary but "complex orthogonal"). We must also indicate the transformation i.e. the correspondence of linear operators. Any linear operator must in the new frame be represented by a matrix, which operates on an a_ℓ thus

$$b_k = \sum_\ell c_{k\ell} a_\ell. \tag{3,4}$$

If according to the aforesaid analogy we transfer the familiar terms of ordinary space to function space, we have to call a wave function, as a_ℓ or b_k, a *vector* and an operator $c_{k\ell}$ a *tensor of the second rank*. This transfer of concepts, which is sometimes a very convenient help, suggests itself particularly in the discrete case, but it is by no means restricted to it. It holds equally well in the original x-frame or any other continuous frame, though the idea of a vector with a continuous array of components is a little more exacting. I beg the reader to reconsider the last two paragraphs of Section 1 with regard to this analogy.

How are we to settle the correspondence between a $c_{k\ell}$-array and a linear operator ϑ in the x-frame? We want the operator $c_{k\ell}$ of (3,4) to correspond to *that* operator ϑ of the x-frame, which *always* turns the wave function which in the x-frame corresponds to a_ℓ into the wave function which in the x-frame corresponds to b_k; "always" means that this must hold for *any* set a_ℓ and the set b_k that is associated with it by (3,4). Or briefly: we want the association between *operand* and *operated* to be frame-invariant. Take ϑ to be a differential operator. Let

$$a_\ell = \int f^*(x, \ell) \varphi(x)\, dx$$

$$b_k = \int f^*(x, k) \vartheta \varphi(x)\, dx$$

We demand

$$\int f^*(x, k) \vartheta \varphi(x)\, dx = \sum_\ell c_{k\ell} \int f^*(x, \ell) \varphi(x)\, dx$$

48 • Chapter 2. Transformation and Interpretation

or

$$\int \left[\vartheta^+ f^*(x, k) - \sum_\ell c_{k\ell} f^*(x, \ell) \right] \varphi(x)\, dx = 0$$

where ϑ^+ is the adjoint. This is to hold for *any* $\varphi(x)$, thus

$$\vartheta^+ f^*(x, k) = \sum_\ell c_{k\ell} f^*(x, \ell)$$

and therefore

$$c_{k\ell} = \int f(x, \ell) \vartheta^+ f^*(x, k)\, dx$$

$$= \int f^*(x, k) \vartheta f(x, \ell)\, dx. \qquad (3,5)$$

It is seen that $c^*_{\ell k} = c_{k\ell}$ if $\vartheta^+ = \vartheta^*$. We shall not enter at the moment into the extension to integral or more general operators ϑ. The fundamental request remains always the same, namely that the operational association of wave functions be independent of the frame. This has an important consequence. If we call product ϑP of the two operators ϑ and P the operation of performing first P and then ϑ, then, according to the said invariance, to the product ϑP obviously corresponds the product, in the same order, of the matrices representing ϑ and P in the new frame. For we must keep in mind that an operator is nothing but the sum total of associations it establishes between wave functions. Now this consideration can be extended to any number of factors; moreover it holds quite obviously also for the sum of any number of summands. The net result is: we may state, that certainly any *polynomial* relation between operators is frame-invariant, and we may presume that the extension to any *analytical* will be allowed, whenever such a one prevails and has a meaning.—The warning is perhaps not superfluous, that nothing of the sort holds for wave functions—nothing except the conservation of linear relations with constant coefficients.

Though it is logically superabundant I should like to add the formal proof for the case of the product of two differential operators. Let in addition to (3,5)

$$g_{\ell m} = \int f^*(x, \ell)\, P\, f(x, m)\, dx. \qquad (3,6)$$

Thus, using the *first* expression for $c_{k\ell}$ in (3,5)

$$\sum_\ell c_{k\ell} g_{\ell m} = \sum_\ell \int f(x, \ell) \vartheta^+ f^*(x, h)\, dx \cdot \int f^*(y, \ell)\, P\, f(y, m)\, dy.$$

Write this as a double integral and draw $\sum_\ell$ under the integral, where then $\sum_\ell f(x, \ell) f^*(y, \ell) = \delta(x, y)$, so that, say, the $\int dy$, when performed first, gives you (*under* the sign $\int dx..$) Pf(x, m). Then throw the ϑ^+ on to this Pf(x, m), giving ϑ Pf(x, m). This proves, that the matrix product on the left represents ϑ P, as we wanted to prove.

For practical application the formula (3,5) or (3,6) is not handy. It is best to procure oneself (either with or without it) the matrices representing x and $\frac{d}{dx}$ (or $-i\frac{d}{dx}$) and, using the invariance of operator relations, to compose other operators from them by matrix multiplication.

4 • The Interpretation of the Wave Function

Thus the wave function is a function of the label ℓ, discrete or continuous according as the frame is discrete or continuous, that is as the functions $f(x, \ell)$, that transform from the x-frame into the frame in question are discrete or continuous with respect to ℓ. The mixed case is possible, but rare. The label ℓ may mean the eigenvalue of an operator or not, the latter case being more general, since, given any orthonormal set $f(x, \ell)$, one can at any time define a Hermitian operator uniquely as the one that has this set for eigenfunctions with arbitrarily chosen real E_ℓ for eigenvalues. The different operators obtained in this way from the *given* set of functions and *any* set of E_ℓ are, generally speaking, functions of each other.—The transforming functions $f(x, \ell)$ are *in the new frame* expressed by $(1, 0, 0, 0, \ldots)$, $(0, 1, 0, 0, \ldots)$, $(0, 0, 1, 0, 0, \ldots)$, etc., these sets going over into Dirac δ-functions, if (or in the domain where) the dependence of ℓ is continuous. It is perfectly possible to contemplate a system in a frame of this kind without any reference to the x-frame. This may seem a rather void scheme, but actually not much more void than the x-frame is hitherto, since the connection with observable facts remains to be indicated anyhow.

The interpretation of the wave function can be given in various equivalent ways. In any case one says that with some, possibly with all, Hermitian operators ϑ is associated a certain experimental device of making an observation on the system that results in determining a real number, part of the characteristic of the present state of the system and called *the value of that observable*. The wave function is supposed to contain nothing less *and nothing more* than a comprehensive description of the state of the system at a given moment with regard to all characteristics of this kind. The wave-function, characterizing the state of the system, and the operator, characterizing the experimental device, together are supposed to give a certain information on the observed value. We shall discuss this connection—which we call the interpretation—in detail.[2] Before doing so, let me emphasize that however accurate, clear and precise this discussion is, it still remains void, before you specify with equal precision the association of experimental devices and linear operators. This is vital, it is a thing that mathematics alone cannot supply, and therefore it is a thing that theoreticians are prone to overlook or at least to set aside as a minor accessory. I am underlining this in advance for the following reason. The interpretation, as you will see, does stipulate *one trait* of this *association* either explicitly or implicitly. And that can easily induce one to believe that it specifies the association completely.

We had best begin with this trait. We spoke of an experimental device that leads to determining a real number. This means that after performing the experiment, taking some pointer readings and executing some computations we write down a number in our record. Now instead of writing this number we could have decided, in all cases of this measurement, to register the third power, the logarithm, the hyperbolic sine or any single valued real function of that number. The experimental device being the same, the physicist would say, that he has performed the same measurement. If the chosen function is monotonic, it is just another way of recording his results, may be more convenient for some purpose. If it is not monotonic, say the square the hyperbolic sine or the ordinary sine, then the registration is incomplete, some of the information is dropped, but it may be an irrelevant part, whose loss is outweighed by the more appropriate form of the relevant part. (An example could be "Brownian displacement," where you may prefer to register the square though this obliterates the sign).

Supposing that with the first form of recording the operator ϑ is associated, what operator shall we associate with any other form, say using the recording function f( )? *We decide that it should be* $f(\vartheta)$. I have made serious attempts to prove that every other assumption would be unsound. I have not succeeded. If anyone notices in the following explication a point where the unsoundness of any alternative assumption emerges, I shall welcome his comment. *The assumption*

---

[2] In all this the *nature* of the system is not yet formally involved. It is connected with the *Hamiltonian* which plays no part here.

is vital. We call it the assumption of correspondence. It is a loan from classical physics. It is suggested by the fact that many of our operators are loans from classical physics, they are functions taken from classical physics, "translated" into operators. In classical physics, if I call the measured length of a rod x, then the observable "square of the length" *has* to be called x^2.

Once this stated, the interpretation can be summarized in one sentence: the *expectation value* of an observable is the inner product of the actual wave function into the function that results from operating on the same by the operator associated with the observable in question. To fix the ideas, take a discontinuous representation, and let the wave function be a_ℓ, the matrix, representing the operator, $\vartheta_{k\ell}$. Then

$$\bar{\vartheta} = \sum_k a_k^* \sum_\ell \vartheta_{k\ell} a_\ell \qquad (4,1)$$

(If the representation were continuous $\sum_k$ would be an integral while $\sum_\ell$ might be some differential operation performed on the wave function.) By expectation value $\bar{\vartheta}$ we mean: it is the mean value to be expected when you make a great many observations with the same experimental device on the system in the same initial state, described by a_ℓ.

Note first that this $\bar{\vartheta}$ is *invariant* according to our explanation of the change of frame. That is alright, but does this contain information on the detailed statistics? It does, in virtue of the "assumption of correspondence." Fix your attention to an interval $\vartheta, \vartheta + \Delta$ of numerical values, and *change your recording*, writing just *1* in your record, when the observed value lies inside this interval, and zero when it does not. This is a non-monotonic function of the observable, so we lose a lot of information; but that is just what we want. For the expectation value of this observable $\overline{f(\vartheta)}$ is the probability (or relative frequency) of finding the observable inside Δ. In principle we can compute it with any desired accuracy. For we can form the matrix ϑ^2, viz. $\vartheta_{k\ell}\vartheta_{\ell m}$, and ϑ^3, viz. $\vartheta_{k\ell}\vartheta_{\ell m}\vartheta_{mn}$ etc.; so we can form any analytical function of ϑ. And we can, of course, approximate our $f(\vartheta)$ arbitrarily well by an analytical function. We are here explaining the principle. Fortunately mathematicians have worked out more expedient methods of putting it into practice.

An interesting and unexpected consequence of our assumptions emerges at once. Let b_ℓ be an eigenfunction of ϑ, with eigenvalue λ_ℓ. It is *obviously* also an eigenfunction of *any* $f(\vartheta)$, with eigenvalue

52 • *Chapter 2. Transformation and Interpretation*

f(λ_ℓ). Hence *our* f(ϑ) just reproduces an eigenfunction b_ℓ of ϑ when λ_ℓ is inside Δ, annihilates it when λ_ℓ is outside. Now consider our

$$\overline{f(\vartheta)} = \sum_k a_k^* \sum_\ell [f(\vartheta)]_{k\ell} a_\ell . \qquad (4,2)$$

Whatever a_ℓ may be, it can be developed with respect to the eigenfunction b_ℓ of ϑ. Hence $\overline{f(\vartheta)} = 0$, *unless* an eigenvalue λ_ℓ of ϑ is inside Δ. Since this holds of *every* interval, our assumptions imply, that the eigenvalues λ_ℓ of ϑ are the only results our "device of observing ϑ" can produce. This shows on the one hand how very far from actual facts our scheme is *in some cases*, while on the other hand it recalls to us from which domain of physics this scheme originated, namely from the observation and theory of line spectra.

It is suggestive to transform a_ℓ to the frame where ϑ is diagonal, its diagonal element $\vartheta_{\ell\ell}$ being then λ_ℓ, with the eigenfunction $0, 0, 0, \ldots, 1, 0, 0, \ldots$, the 1 standing in the ℓth place (you might write $e^{i\alpha}$, with α real, instead of 1, that makes no difference). Let us call a_ℓ, in this frame, b_ℓ; to use the same letter as just before, is a little, but not very inconsistent, and quite a good reminder of the fact, that now "an isolated coefficient b_ℓ, normalized" *is* an eigenfunction of ϑ. Then we have

$$\bar{\vartheta} = \sum_\ell b_\ell^* \lambda_\ell b_\ell . \qquad (4,3)$$

What is f(ϑ) in this frame? It is also diagonal, the λ_ℓ inside Δ being replaced by one's, those outside Δ by zero's. Hence

$$\overline{f(\vartheta)} = \sum_{\lambda_\ell \text{in} \Delta} b_\ell^* b_\ell . \qquad (4,4)$$

In words: the sum (or integral) over Δ of the absolute value of the wave-function, expressed as a function of the eigenvalue of ϑ, is the probability of finding the corresponding observable inside Δ.

The preceding sentence, by itself alone, embodies the *whole* interpretation, it comprises (4,1) *and* the assumption of correspondence. It says briefly that $b_\ell^* b_\ell$ (in the frame where ϑ is diagonal!) is the relative frequency of observing λ_ℓ—with a slight complication of speech, unimportant at the moment, when several λ_ℓ coincide (degeneracy). From this you draw (4,3) as a trivial consequence; in addition—if you choose to modify your recording using *any* function g( )—the mean value of these g's, always to be formed of the observed value that can only be one of the λ's, is obviously and trivially

$$\sum_\ell b_\ell^* g(\lambda_\ell) b_\ell ;$$

but since the operator $g(\vartheta)$ is also diagonal and has the eigenvalues $g(\lambda_\ell)$, this can obviously be written:

$$\sum_\ell b_\ell^* \sum_k [g(\vartheta)]_{\ell k} b_k .$$

The trivial fact, that this according to the new form of basic statement *is* our expectation value for the modified form of recording, can now easily be transformed to any frame, since the above expression is written in invariant form (and the *observed* mean can have nothing to do with the frame). This gives (4,1) if you choose $g(\vartheta) = \vartheta$, and gives the "assumption of correspondence," since you may choose any other g( ), and the whole deduction holds for *any* operator.

Summarizing, the interpretation may be founded axiomatically in two ways, to wit

(i) correspondence axiom + (expectation value = inner product).
(ii) statistical axiom in the diagonal frame of ϑ.

I believe the two to be entirely equivalent. The second is shorter, but decidedly more artificial. You have to swallow a greater lump at a time. You have to assume explicitely that the system can never be found in a non-eigenstate, when this quantity is measured! The correspondence axiom seems very natural. The second part of (i) is less plausible. It will seem more natural when we come to introduce the Hamiltonian operator which transforms the wave-function as time goes on, and come to think of any operator as a possible *perturbing addition* to the Hamiltonian. This perturbation produces within a brief interval of time an addition of the form $i\vartheta\psi$ to ψ, which, on forming $|\psi|^2$ would produce exactly terms of the form $\psi \cdot \vartheta\psi$. There seems to be a connection between the *transitions* caused by a perturbing operator ϑ and the expectation value of the observable associated with ϑ. For all I know, the problem has not been put or attacked in this way, though the answer may be implicitly contained under quite different form in other considerations.

54 • Chapter 2. Transformation and Interpretation

5 • Illustrations

I wish to give examples to illustrate the transformation theory, examples of a general kind, that will bring out relevant general facts.

The quantum transformation of outstanding importance is the Fourier-integral-transformation which leads from the x-frame to another continuous frame, the p-frame. Remember that a transformation is given by a function f(x, ℓ) of two variables, which are essentially the eigenvalues of two *operators* x and ℓ, which are "diagonal" in the two frames, respectively, and which are the arguments of the wave function in the two frames, respectively; more precisely the wave function is in every frame a function of the eigenvalues of *that* operator that is diagonal in that frame. In the present case we write p for ℓ, which as I said is here also a continuous variable, and the transformation function is

$$f(x, p) = \frac{1}{\sqrt{2\pi}} e^{-ixp}. \tag{5,1}$$

Thus if $\psi(x)$ is the wave function in the x-frame, it reads in the p-frame

$$\varphi(p) = \frac{1}{\sqrt{2\pi}} \int e^{-ixp} \psi(x) \, dx. \tag{5,2}$$

It follows

$$i\frac{d}{dp}\varphi(p) = \frac{1}{\sqrt{2\pi}} \int e^{-ixp} x\psi(x) \, dx. \tag{5,3}$$

From the Fourier inversion theorem the inverse transformation is

$$\psi(x) = \frac{1}{\sqrt{2\pi}} \int e^{ixp} \varphi(p) \, dp \tag{5,4}$$

thus

$$-i\frac{d}{dx}\psi(x) = \frac{1}{\sqrt{2\pi}} \int e^{ixp} p\varphi(p) \, dp. \tag{5,5}$$

Using not the general transformation formula for operators (of which I told you it was not expedient) but the guiding principle that operators associated in the two frames are such as turn associated functions again into associated functions, we state the association

$$\begin{array}{cc} \text{x-frame} & \text{p-frame} \\ x \to & i\dfrac{d}{dp} \\ -i\dfrac{d}{dx} \to & p \end{array} \qquad (5,6)$$

of which the last is particularly interesting, since it tells us that p means in the more familiar x-frame just the differentiator, multiplied by −i (which makes it self-adjoint; so does i; clearly the factor has to be i in one case, −i in the other). Moreover the "inverse" transformation function e^{+ixp}, as a function of x, is an eigenfunction of the second operator

$$-i\frac{d}{dx} e^{ixp} = p e^{ixp}, \qquad (5,7)$$

eigenvalue p (with the *vice versa* about e^{-ixp} as a function of p). Keeping to the first statement, one desires the *transformed* of e^{ixp} to be an eigenfunction of "the multiplier" p, i.e. a δ-function. What do we get for it?

$$\frac{1}{\sqrt{2\pi}} \int e^{-ixp} \cdot e^{ixp'} \, dx = \frac{1}{\sqrt{2\pi}} \int e^{-ix(p-p')} \, dx = \delta(p, p') \qquad (5,8)$$

The integral is divergent. It is the special form of the bilinear series of orthonormal functions by which the δ-function is always expressible. The last equation means to say that if, *under* the sign of integration, you multiply by *any* f(p) dp, and integrate, *under* the sign of integration with respect to p, you get f(p′)—by Fourier's double integral-theorem.

The fact that the operations x and $-i\frac{d}{dx}$ do not commute, but

$$-i\frac{d}{dx} x - x\left(-i\frac{d}{dx}\right) = -i \qquad (5,9)$$

is sometimes expressed by

$$px - xp = -i \qquad (5,10)$$

which is convenient, but not very appropriate since it mixes up notations that naturally refer to different frames. We call a couple of operators with this commutation relation canonically conjugate. But the coupling is not unique, since the *operator* p + f(x) has the same commutation relation with x, and x + g(p) the same with p (but you may only change *one* of them in this manner, if they are to remain conjugate!). A pair of operators of singular interest—though *not* Hermitian or self-adjoint, but mutually adjoint, is:

$$\xi = x - ip = x - \frac{d}{dx}$$
$$\eta = x + ip = x + \frac{d}{dx}.$$
(5,11)

One has

$$\eta\xi = x^2 + p^2 + 1, \qquad \frac{1}{2}(\eta\xi + \xi\eta) = x^2 + p^2$$
$$\xi\eta = x^2 + p^2 - 1, \qquad \frac{1}{2}(\eta\xi - \xi\eta) = 1.$$
(5,12)

If you express them in the p-frame

$$\xi = x - \frac{d}{dx} \to i\frac{d}{dp} - ip = -i\left(p - \frac{d}{dp}\right)$$
$$\eta = x + \frac{d}{dx} \to i\frac{d}{dp} + ip = i\left(p + \frac{d}{dp}\right).$$
(5,13)

What functions are reproduced by them, up to a constant?

$$\xi f = af \qquad \left(x - \frac{d}{dx}\right)f = af \qquad \frac{df}{dx} = (x - a)f \qquad f = e^{\frac{1}{2}(x-a)}$$
$$\eta f = af \qquad \left(x + \frac{d}{dx}\right)f = af \qquad \frac{df}{dx} = (a - x)f \qquad f = e^{-\frac{1}{2}(x-a)}$$

(These are *not* eigenfunctions and eigenvalues, the operators being not self-adjoint.) Envisage the 2nd for a = 0. Call it $f_0 = e^{-x^2/2}$. It is annihilated by η. Hence also by $\xi\eta$. Thus

$$\xi\eta f_0 = (x^2 + p^2 - 1)f_0 = 0$$
(5,14)

This gives us one eigenfunction of the Hermitian $x^2 + p^2 = H$, say

$$\xi|Hf_0 = f_0.$$

Now notice that

$$\xi\eta\xi = \xi(H + 1) = (H - 1)\xi \qquad \text{thus} \qquad H\xi - \xi H = 2\xi. \tag{5,15}$$

Further

$$\xi H f_0 = \xi f_0 = (H\xi - 2\xi) f_0$$
$$H\xi f_0 = 3\xi f_0.$$

This can be continued; one gets: $\xi^n f_0 = f_n$, say, is an eigenfunction of the Hermitian H, eigenvalue $2n + 1$. It is a polynomial of degree n (even or odd according as n is) times $e^{-x^2/2}$. They are called Hermite-polynomials and Hermite-orthogonal functions (not yet normalized).

At the moment the more interesting thing is: what does this p mean? We have become used to saying that $-i\frac{d}{dx}$ is "of course" the momentum conjugate—but what suggests this association? That this question turns up—and similar question turn up again and again—is an illustration to what I said about the "mathematical theory of interpretation" being a void scheme, which needs to be supplemented by establishing the association between experimental devices and linear operators. This is the first relevant case to the point. I said the establishment of this association is outside the domain of the mathematical apparatus. So you must not expect anything like a mathematical deduction. But on the other hand we need something more general than a handbook of experimental physics describing all the experimental devices in the world that have ever been put into practice (and assigning an operator to each of them). Now, if we say $-i\frac{\partial}{\partial x}$ means (or is associated with) the *momentum*, then every physicist has at least a general idea of some appropriate measuring devices for the momentum. We may some time have to go into more details about it, but for the moment we can requiesce in this. But now, why should it mean the momentum?

There are two largely independent sources to suggest this, namely i) the conjunction of Planck's quantum hypothesis and Special Relativity ii) Hamilton's analogy between optics and mechanics. The first idea was implicitly contained in de Broglie's famous thesis of 1925, though the notions of wave functions and linear operators were still alien to it. This connection was only revealed, when, a year or two later, the intimate relation with Hamiltonian mechanics occurred to me. The interesting thing about it, was, that the second pillar, Planck's fundamental "hypothesis of discreteness," was no longer needed; this does not mean that it was dropped, but it came out by itself in the form, that certain operators, constructed on the basis of the Hamiltonian analogy, have discrete eigenvalues.

I shall not explain historically the two different ways of attack, but rather give a synthesis.

What is p? Its eigenfunction in the x-frame is

58 • *Chapter 2. Transformation and Interpretation*

$$e^{ixp} \quad \text{(as a function of x)},$$

p being the eigenvalue parameter. This is a periodic function, a "sine-wave," p being 2π times the reciprocal wave length, called the wave-number. But mind you, at the moment we are still faced only with a periodic function of x, we call wave-length the distance between consecutive crests; the notion of time is not involved. But we cannot help *thinking* of waves and *remembering* that with e.g. a light-wave in 3-dimensional space the wave-number has 3 components $(\alpha/\lambda, \beta/\lambda, \gamma/\lambda)$ and constitutes the 3 spatial components of a 4-vector whose time component is the frequency ν. About this *frequency* we know that according to Planck (1900) and Einstein (1905) it determines an amount of energy $h\nu$, and about energy we know that it is also the 4th component of a 4-vector, whose space components are those of linear momentum. This suggests to extend the Planck-Einstein hypothesis and regard the wave-number connected in the same way with the linear momentum belonging to the said amount of energy. Which amount of energy, in our case? Well we started from a coordinate x. If we regard it as the abscissa of a mass-point moving along some straight line, we shall obviously try to take the momentum and energy of this mass-point

$$\frac{mv}{\sqrt{1-\beta^2}}, \quad \frac{mc^2}{\sqrt{1-\beta^2}} \cdot \frac{1}{c} \quad \left(\beta = \frac{v}{c}\right)$$

(We have added a factor $\frac{1}{c}$ to get the dimensions homogeneous).
The other 4-vector is

$$\frac{1}{\lambda}, \quad \frac{\nu}{c} \quad \text{(again } \tfrac{1}{c} \text{ for dimensions)}$$

Thus

$$\frac{mv}{\sqrt{1-\beta^2}} = \frac{h}{\lambda} \leftarrow \frac{mc^2}{\sqrt{1-\beta^2}} \cdot \frac{1}{c} = \frac{h\nu}{c}.$$

(This by the way is not at all a quantization of energy. It means attaching a frequency to the amount of energy the mass-point possesses.)
Thus our

$$p = \frac{2\pi}{\lambda} = \frac{2\pi}{h} \frac{mv}{\sqrt{1-\beta^2}}. \tag{5,16}$$

(We see, that if we want p itself to "mean" the momentum, we have to change its definition by a constant, but that is a minor point; we can always choose our units so as to make $h = 2\pi$, *and we usually shall*.)

By introducing the notion of frequency we have inadvertently introduced the notion of time and of velocity of propagation, say u:

$$u = \lambda \nu = \frac{\frac{mc^2}{\sqrt{1-\beta^2}}}{\frac{mv}{\sqrt{1-\beta^2}}} = \frac{c^2}{v}. \tag{5,17}$$

We are taken aback. We find all our good intentions foiled. We might have expected u = v, instead of which we get, *horribile dictu*, u > c, since v ≤ c. I wonder whether this was for de Broglie the reason for not speaking explicitly of the wave equation? Anyhow the solution of this apparent impasse is now well known. Form the group velocity

$$v_g = d\nu/d(1/\lambda) = d\left(mc^2/\sqrt{1-\beta^2}\right) / d\left(mv/\sqrt{1-\beta^2}\right)$$
$$= v \quad \text{(known from the Hamiltonian equations of motion)} \tag{5,18}$$

This order of ideas has surreptitiously introduced the new assumption, that wave functions change with time, and that in a very well defined fashion. Our wave function

$$\left. \begin{array}{c} \psi = e^{ixp} \\ \text{for which} \\ -i\frac{d}{dx}\psi = p\psi \end{array} \right\} \tag{5,19}$$

is supposed to change in time according to

$$\frac{d}{dt}\psi = 2\pi i\nu\psi \quad \text{(this means "with frequency } \nu\text{")}$$

where ν is a certain function of p, that is it depends on the way ψ depends on x. In what way? In order to avoid changes in the notation, let us use units such that $h = 2\pi$ and $c = 1$.
    Then

$$2\pi\nu = \frac{m}{1-v^2} \qquad p = \frac{mv}{\sqrt{1-v^2}}$$

and one knows that then

$$2\pi\nu = \sqrt{m^2 + p^2} \tag{5,20}$$

(The relativistic invariant $m^2 = 4\pi^2\nu^2 - p^2$; $2\pi\nu$ "for energy"). Thus

$$-i\frac{d\psi}{dt} = \sqrt{m^2 + p^2}\,\psi. \tag{5,21}$$

This holds for the eigenfunction e^{ixp}, for which $-i\frac{d}{dx}\psi = p\psi$. So for every one with another constant p. To give it the same expression in all cases write

$$-i\frac{d\psi}{dt} = \sqrt{m^2 - \frac{d^2}{dx^2}}\,\psi \qquad (5,22)$$

We *assume*, that it shall hold for any linear aggregate of them (superposition of "waves")—but then for *any* function ψ.

The square-root is a *linear* operator. Mind you, that it simply is $\vartheta(a\varphi + b\psi) = a\vartheta\varphi + b\vartheta\psi$. That is no new assumption—I beg to distinguish this from the assumption of superposition.—It has a queer form, it includes space derivatives up to any order. One suspects that it its not a differential equation at all, but involves finite differences, just as e.g.

$$e^{a\frac{d}{dx}}f(x) \equiv f(x+a) \qquad (5,23)$$

But this is, I think, not so. Any solution of (5,22) must fulfil the iterated equation

$$-\frac{\partial^2\psi}{\partial t^2} = \left(m^2 - \frac{\partial^2}{\partial x^2}\right)\psi. \qquad (5,24)$$

which is the simplest case of what most people call the Klein-Gordon equation. The *vice-versa cannot* be true, because in (5,24) we may choose the initial values of ψ and $\frac{\partial\psi}{\partial t}$ arbitrarily, while in (5,22) the latter are prescribed by the former.

I wish to prove that every solution of (5,24) is *essentially* the sum of a solution of (5,22) and of (5,22') ((5,22) taken with the opposite sign of the square root). We may decompose the operator of (5,24) thus

$$\left(\sqrt{m^2 - \frac{\partial^2}{\partial x^2}} - i\frac{\partial}{\partial t}\right)\left(\sqrt{m^2 - \frac{\partial^2}{\partial x^2}} + i\frac{\partial}{\partial t}\right)\psi = 0$$

Hence if we put

$$\left(\sqrt{m^2 - \frac{\partial^2}{\partial x^2}} + i\frac{\partial}{\partial t}\right)\psi = \varphi$$

$$\left(\sqrt{m^2 - \frac{\partial^2}{\partial x^2}} - i\frac{\partial}{\partial t}\right)\varphi = 0$$

and if we put

$$\left(\sqrt{m^2 - \frac{\partial^2}{\partial x^2}} - i\frac{\partial}{\partial t}\right)\psi = \chi$$

$$\left(\sqrt{m^2 - \frac{\partial^2}{\partial x^2}} + i\frac{\partial}{\partial t}\right)\chi = 0$$

It follows

$$2i\frac{\partial \psi}{\partial t} = \varphi - \chi$$

$$\psi = \Phi + X + f(x)$$

hence with

$$\frac{1}{2i}\int^t \varphi\, dt = \Phi, \qquad -\frac{1}{2i}\int^t \chi\, dt = X$$

also solutions of (5,22) and (5,22′) and

$$\frac{d^2 f}{dx^2} - m^2 f = 0, \qquad f = e^{\pm mx}.$$

To this addition—which would be a little more general, if we had 3 dimensions and the Δ-operator instead of just the second derivative—I referred by "essentially." One must *not* say, oh well, of course

$$\sqrt{m^2 - \frac{d^2}{dx^2}}\, f = 0,$$

because that series diverges. Anyhow this f can never turn up in a wave-function on account of its behaviour at infinity, at the positive or negative end.

It is interesting that the difficulty with the ambiguous sign of the energy springs from relativity theory quite generally; it is not a result of the special Dirac spin theory.

Let us now attend to a rather trivial matter. Some physicists drag physical constants all along their mathematical formulae, because they are afraid of losing them. We have put $h = 2\pi$ and $c = 1$. Let us assume that we had done work, putting $m = 1$, and arrived at

$$-\frac{\partial^2 \psi}{\partial t^2} = \left(1 - \frac{\partial^2}{\partial x^2}\right)\psi;$$

now we wish to get the equation in ordinary units. How manage? We just stick in factors $h/2\pi$, c and m, so as to put the dimensions in order.

That is all. In our case the first thing to stick in is a $1/c^2$ on the left. What about the 1? It has to be the square of a reciprocal length. mc/h is one—the only one! So we *must* stick in $(2\pi mc/h)^2$—for this was 1 in the units chosen.

Now I come to an important distinction—a new assumption, if you please. We regard the association of the Fourier transformed variable p with the momentum as quite general in every respect. By "in every respect" I mean i) it has nothing to do with the equation of time change, and ii) whatever x is, its Fourier transform shall be associated with the partial derivative of the kinetic energy with respect to x. As against this, the special equation of time change which we have *now* derived and which reads, in the units, $h = 2\pi$,

$$-i\frac{\partial \psi}{\partial t} = \text{(kinetic energy expressed by the momenta) operating on } \psi,$$

is regarded as controlling "a force-free mass-point." In the general case of any system it will be the same, but one must take the *total* energy expressed by the momenta. Moreover for *slow* motion one may now waive relativity and write

$$-i\frac{\partial \psi}{\partial t} = m\left(1 + \frac{1}{2m^2}p^2\right) + \text{Potential energy}$$

$$= m + \frac{1}{2m}p^2 + \text{Potential energy} = H(p)$$

(operating on ψ)

and one can let m be absorbed in the Potential energy, whose zero level is meaningless anyhow.

An example which has acquired prominence far beyond its original meaning is the "Planck-oscillator":

$$H = \frac{1}{2m}p^2 + \frac{1}{2}f^2 x^2$$

We allow ourselves for general discussion to put $m = \frac{1}{2}$, $f^2 = 2$. (Since c no longer turns up we may liquidate $c = 1$.) Then we have $H = p^2 + x^2$, of which we know the eigenfunction, viz. $\xi^n e^{-x^2/2}$, eigenvalue $2n + 1$. What does an eigenfunction of the Hamiltonian mean in general? Two things. First it has a very simple dependence on time

$$-i\frac{\partial \psi}{\partial t} = H\psi = \lambda \psi, \qquad \psi = e^{i\lambda t} \cdot \text{eigenfunction}[3].$$

---

[3] This has the consequence that all statistics are independent of time. The state does not change. Nothing happens.

Secondly: if the wave function is one of them, the energy has the value λ. We now proceed to give a second example of transformation, viz. to the eigenfunctions of the oscillator-H. Please note however that we are now turning back to our early time-free aspect. We can make this transformation for *any* x, whether its time dependence is controlled by the oscillator equation or not.

6 • (Harmonic Oscillators)

We shall first put together some properties of the Hermite functions. The way we obtained them, to wit,

$$h_n(x) = \xi^n h_0(x), \qquad \xi = x - \frac{d}{dx}, \qquad h_0(x) = e^{-x^2/2}$$

is not the most convenient, because x and $\frac{d}{dx}$ do not commute, so that e.g.

$$\xi^2 \neq x^2 - 2x\frac{d}{dx} + \frac{d^2}{dx^2}.$$

We can replace it by a simpler iteration. The h_n is obviously a polynomial, degree n, times $e^{-x^2/2}$. Say

$$h_n(x) = H_n(x)e^{-x^2/2} = \xi^n h_0(x)$$

$$H_n(x) = e^{x^2/2} h_n(x) = e^{x^2/2} \xi h_{n-1}(x)$$

$$= e^{x^2/2} \xi e^{-x^2/2} H_{n-1}(x)$$

We analyse this *operation* quite in general

$$e^{x^2/2}\left(x - \frac{d}{dx}\right)e^{-x^2/2} = x + x - \frac{d}{dx} = 2x - \frac{d}{dx};$$

compare it with

$$e^{x^2}\left(-\frac{d}{dx}\right)e^{-x^2} = 2x - \frac{d}{dx};$$

this is the *same*, and that quite irrespective of the function on which it operates. Hence

$$H_n(x) = e^{x^2}\left(-\frac{d}{dx}\right)^n e^{-x^2} 1 \qquad (H_0(x) = 1 \text{ as you see!})$$

We shall use the forthwith. Just a short digression. ξ has a twin brother, to wit $\eta = x + \frac{d}{dx}$. What about the *operator*:

$$e^{x^2/2}\left(x + \frac{d}{dx}\right)e^{-x^2/2} = x - x + \frac{d}{dx} = \frac{d}{dx}$$

$$e^{-x^2/2}\left(x + \frac{d}{dx}\right)e^{x^2/2} = x + x + \frac{d}{dx} = 2x + \frac{d}{dx} = e^{-x^2}\frac{d}{dx}e^{x^2}$$

These are sometimes useful transformations. End of digression.

We use an auxiliary variable t (which has nothing to do with time) and form

$$\sum_0^\infty \frac{H_n(x)}{n!}t^n = \sum_0^\infty e^{x^2}(-t)^n \frac{1}{n!}\frac{d^n}{dx^n}\left(e^{-x^2}\right) = e^{x^2-(x-t)^2}$$

This is a generating function for the Hermite polynomials, and, when multiplied by $e^{-x^2/2}$, for the (not yet normalized) Hermite functions. We use it to show that they are orthogonal (as they must, being the eigenfunctions of a Sturm-Liouville-problem) and to normalize them:

$$e^{x^2/2-(x-t)^2} \cdot e^{x^2/2-(x-s)^2} = \sum_0^\infty \sum_0^\infty \frac{h_n(x)h_m(x)}{n!\,m!}t^n s^m$$

$$\int_{-\infty}^{+\infty} e^{-x^2+2x(t+s)-t^2-s^2}\,dx = e^{2ts}\sqrt{\pi} = \sqrt{\pi}\sum_0^\infty \frac{t^n s^n 2^n}{n!}.$$

Thus

$$\int_{-\infty}^{+\infty} h_n(x)h_m(x)\,dx = 2^n n!\sqrt{\pi}\,\delta_{nm},$$

which shows the orthogonality and that the $h_n(x)$ used hitherto will be normalized by the factor $(2^n n!\sqrt{\pi})^{-\frac{1}{2}}$. For the moment we continue to use them as nature has given them to us.

We used the generatrix also to find their Fourier-transform. We must use it, of course, for the $h_m(x)$. I find it a little more convenient to replace t by $\frac{1}{2}$t.

$$e^{-x^2/2+xt-t^2/4} = \sum_0^\infty \frac{h_n(x)}{2^n n!}t^n$$

$$\int_{-\infty}^{+\infty} e^{-x^2/2+xt-ipx}\,dx = \int_{-\infty}^{+\infty} e^{-\frac{1}{2}[x^2+2(t-ip)x+(t-ip)^2]+\frac{1}{2}(t-ip)^2}\,dx$$

$$= \sqrt{2\pi}\,e^{\frac{1}{2}(t-ip)^2}$$

So we have

$$\sqrt{2\pi}\, e^{\frac{1}{2}t^2 - itp - \frac{1}{2}p^2 - t^2/4} = \sum_0^\infty \frac{t^n}{2^n n!} \int h_n(x) e^{-ipx}\, dx.$$

$$\sqrt{2\pi}\, e^{-\frac{1}{2}p^2 - itp + t^2/4} = \sum_0^\infty \frac{t^n}{2^n n!} \int h_n(x) e^{-ipx}\, dx$$

$$= \sqrt{2\pi} \sum_0^\infty \frac{(-it)^n}{2^n n!} h_n(p)$$

Thus

$$h_n(p) = i^n \frac{1}{\sqrt{2\pi}} \int h_n(x) e^{-ipx}\, dx$$

This is a neat result, well known but little emphasized. It is easy to show, that if a function is proportional to its own Fourier transform, the factor must be $\sqrt[4]{1}$. It is also easy to show, that if for some function f holds

$$f = i^k N f$$

that then

$$\xi f = i^{k+1} N \xi f, \qquad \eta f = i^{k-1} N \eta f.$$

I leave this to you. You must not believe that the Hermite functions are the only ones to have this property. You need only add up *any* function and its Fourier-transform to get a Fourier self-reciprocal. The equation is a singular integral equation with four eigenvalues only. It splits Hilbert space into four linear mutually orthogonal sub-spaces only, or, splits functions into four classes. Any function is a sum of four "class-function." It is a splitting into 2×2, the first being simply the splitting into odd and even functions. The even ones split according to eigenvalues ± 1, the odd ones have $\pm i$.

But let us return to our transformation. We wish to know how x and p transform. What do these continuous variables look like in a discrete representation? We need certain recurrences (we have to express xh and h' by the h's themselves). We get them most conveniently from our original definitions:

Chapter 2. Transformation and Interpretation

$$h_{n+1} = \xi h_n = xh_n - h'_n$$
$$\eta h_{n+1} = (2n+2)h_n \quad \text{or}$$
$$\eta h_n = 2nh_{n-1}$$
$$xh_n + h'_n = 2nh_{n-1}$$
$$xh_n - h'_n = h_{n+1}$$
$$\frac{\sqrt{2n}}{\sqrt{2n}}\left\{\begin{array}{c} xh_n \\ h'_n \end{array}\right\} = nh_{n-1} \pm \frac{1}{2}h_{n+1} \cdot \frac{\sqrt{4(n+1)n}}{\sqrt{4(n+1)n}}$$

Moreover:

$$\eta\xi = H + 1$$
$$\eta\xi h_n = (2n+2)h_n$$

From now on we wish to use normalized h's for convenience. Remember the normalizing factor $\dfrac{1}{\sqrt{2^n n! \sqrt{\pi}}}$

For the moment we *underline* the normalized h's!

$$\sqrt{2n}\left\{\begin{array}{c} x\underline{h}_n \\ \underline{h}'_n \end{array}\right\} = n\underline{h}_{n-1} \pm \underline{h}_{n+1}\sqrt{n(n+1)}$$

$$\left\{\begin{array}{c} x\underline{h}_n \\ \underline{h}'_n \end{array}\right\} = \sqrt{\frac{n}{2}}\underline{h}_{n-1} \pm \sqrt{\frac{n+1}{2}}\underline{h}_{n+1}$$

But we shall from now on drop underlining and make it a standing order.

Now if you have a function $\varphi(x) = \sum a_k h_k$ expressed by the normalized functions, then

$$x\varphi = \sum_{k=0}^{\infty}\left(\sqrt{\frac{k}{2}}a_k h_{k-1} + \sqrt{\frac{k+1}{2}}a_k h_{k+1}\right)$$

$$= \sum_{k=0}^{\infty}\left(\sqrt{\frac{k+1}{2}}a_{k+1} + \sqrt{\frac{k}{2}}a_{k-1}\right)h_k$$

$$= \sum_{k=0}^{\infty} b_k h_k$$

$$b_k = \sqrt{\frac{k+1}{2}}a_{k+1} + \sqrt{\frac{k}{2}}a_{k-1}$$

$$= \sum_i x_{ki} a_i.$$

Hence

$$x_{k,k+1} = \sqrt{\frac{k+1}{2}} \qquad x_{k,k-1} = \sqrt{\frac{k}{2}}$$

The same way (remember $p = -i\frac{d}{dx}$):

$$p_{k,k+1} = -i\sqrt{\frac{k+1}{2}} \qquad p_{k,k-1} = i\sqrt{\frac{k}{2}}.$$

In both cases, the second formula can be replaced by the statement that the matrix is Hermitian. These matrices were first constructed at Göttingen before the functions $h_n(x)$ were known to have anything to do with the matter. They were ingeniously constructed so as to make the matrices

(i) $x^2 + p^2$ diagonal, and
(ii) $px - xp$ diagonal and a multiple of 1.

Let us see how that comes about:
(the $k + 1$ and $k - 1$ in the x-terms of (i) are summation indices)

(i) $\quad x_{k,k+1}x_{k+1,k} + x_{k,k-1}x_{k-1,k} + p_{k,k+1}p_{k+1,k} + p_{k,k-1}p_{k-1,k}$

$$= \frac{k+1}{2} + \frac{k}{2} + \frac{k+1}{2} + \frac{k}{2}$$

$$= 2k + 1$$

(ii) $\quad p_{k,k+1}x_{k+1,k} + p_{k,k-1}x_{k-1,k} - x_{k,k+1}p_{k+1,k} - x_{k,k-1}p_{k-1,k}$

$$= -i\frac{k+1}{2} + i\frac{k}{2} + -i\frac{k+1}{2} + i\frac{k}{2}$$

$$= -i$$

(With a finite matrix, the latter would be impossible, because there a commutator has *trace* zero:

$$\sum_k (c_{ik}g_{k\ell} - g_{ik}c_{k\ell}).$$

If this is summed over $i = \ell$, you get zero; another remark: the commutator of two Hermitian matrices is necessarily, *qua* commutator, skew Hermitian).

What do the eigenfunctions of x look like in our frame? We had just before written out the b_k that represent $x\varphi$, expressed by the a_k

68 • *Chapter 2. Transformation and Interpretation*

representing φ:

$$b_k = \sqrt{\frac{k+1}{2}}\, a_{k+1} + \sqrt{\frac{k}{2}}\, a_{k-1} = \lambda a_k$$

These recurrence equations resemble those of normalized Hermite functions "as one egg resembles the other," x there being replaced by λ here. If you embarked on solving them you would, of course, not get the normalized orthogonal functions of λ, but you would get what one might call the "normalized polynomials," i.e. the functions, multiplied by a *constant* (meaning here *independent of* the *index*). The exponential you would have to add for normalization. (Small wonder!). So to repeat it, in our present frame

$$h_0(x),\ h_1(x),\ h_2(x),\ \ldots,\ h_n(x),\ \ldots$$

is, *for fixed* x, *one* eigenfunction of the operator x, namely that belonging to the numerical value x. What about p? Its eigenvalue problem reads

$$P_{k,\ell}\, a_\ell = \lambda a_k$$

$$-i\sqrt{\frac{k+1}{2}}\, a_{k+1} + i\sqrt{\frac{k}{2}}\, a_{k-1} = \lambda a_k$$

substitute $a_k = i^k S_k$, then you get

$$\sqrt{\frac{k+1}{2}}\, S_{k+1} + \sqrt{\frac{k}{2}}\, S_{k-1} = \lambda S_k \qquad \text{the same as before.}$$

So you would make out that

$$h_0(p),\ i^{+1} h_1(p),\ i^{+2} h_2(p),\ \ldots,\ i^{+n} h_n(p),\ \ldots$$

is *the* eigenfunction belonging to eigenvalue p. Is that correct? If so, this line of coefficients must be the development coefficients of e^{+ipx} in our frame; they read

$$a_k = \frac{1}{\sqrt{2\pi}} \int e^{ipx} h_k(x)\, dx = i^n h_n(p)\,.$$

So this is correct.

What seems to me a rather amusing analytical fact is this. If you form the series

$$\sum_{n=0}^{\infty} h_n(x)h_n(\lambda) = \delta(x,\lambda) \qquad (a)$$

it is a very improper divergent thing representing the Dirac-δ "in a way." The series

$$\sum_{n=0}^{\infty} i^n h_n(x) h_n(p) = e^{ipx} \qquad (b)$$

is obviously entirely respectable and represents an entirely respectable function. (The latter can still be decomposed into real and imaginary parts.)

The state of affairs reminds one very much of an alternating series that does not converge absolutely. It is just one degree more involved. One wonders, would (a) converge with alternating signs?

The x- and p-matrices fill, each of them, two "side-diagonals," these next to the principal diagonal. One can form simpler matrices from them, viz.

$$\xi = x - \frac{d}{dx} = x - ip \quad \text{and} \quad \eta = x + \frac{d}{dx} = x + ip$$

$$\xi_{k,k+1} = 0 \qquad \xi_{k,k-1} = 2\sqrt{\frac{k}{2}}$$

$$\eta_{k,k+1} = 2\sqrt{\frac{k+1}{2}} \qquad \eta_{k,k-1} = 0$$

Let us form their "eigenfunctions," even though they are *not* Hermitian; we take $\frac{1}{\sqrt{2}}\xi$, $\frac{1}{\sqrt{2}}\eta$; for the first we have

$$\frac{1}{\sqrt{2}}\xi_{k,k-1}a_{k-1} = \lambda a_k \qquad \frac{1}{\sqrt{2}}\eta_{k,k+1}b_{k+1} = \lambda b_k$$

$$a_k = \frac{\sqrt{k}}{\lambda}a_{k-1} \qquad b_{k+1} = \frac{\lambda}{\sqrt{k+1}}b_k$$

obviously

$$a_n = \sqrt{n!}\,\lambda^{-n} \qquad b_n = \frac{\lambda^n}{\sqrt{n!}}$$

(apart from a constant)

While the a_n-function is rather odd and cannot be normalized, the b_n-function can, for its square $\frac{\lambda^{2n}}{n!}$ gives, on summation, e^{λ^2}; hence the normalized b_n reads

$$\underline{b}_n = \frac{\lambda^n e^{-\lambda^2/2}}{\sqrt{n!}}.$$

If η were an observable (which it is not, because it is not Hermitian) and the system were an oscillator (which it might be, but it has not been assumed), then

$$|b_n|^2 = \frac{\lambda^{2n} e^{-\lambda^2}}{n!}$$

would be the probability of the oscillator being in the nth level, when η^2 has the value λ^2; and the amusing thing is, that this is Poisson's formula! As if a large amount of energy were distributed in *quanta* at random over a large number of oscillators, λ^2 being the average number any one of them gets.

One gains the impression that some non-Hermitian operators, though they do not represent observables, yet have a certain statistical significance.

We turn back to more general considerations.

7 • (Interpreting the Wave Function)

The wave function is supposed to contain a lot of information, indeed all the information we have about the state of a physical system. I have explained the simple mathematical procedure by which this information is extracted from the wave-function-information about the results of experimental devices applied to the system. We have as yet said nothing on the inverse, to wit how do we come to know the wave function? How are, inversely, the results of "measurements" translated into a wave-function? Quite obviously the same association of linear operators and measuring devices must play a part in this—so we shall take that association for granted, though it remains the most delicate point, not to say the blind spot, of the theory, which *cannot* be filled in by pure mathematics.

In general the state of the system is not such that from it the outcome of a particular device can be unequivocally inferred, not unless the wave-function happens to be an eigenfunction or the operator associated with the measuring device in question. Otherwise the information is only statistical; the *state* does *not* determine the outcome;

(Interpreting the Wave Function) • 71

if you bring your system again and again into the same state,[4] then apply your device, you get sometimes this, sometimes that result from it. But on the other hand we said, the measuring device determines a real number, part of the characteristics of the state of the system. But how can that be, if we get once this and once that and once a third etc. "characteristic number"? Well, we have to acquiesce in this, that *in the state in question* the number determined by the *physical procedure in question* is not a characteristic of the system. Are we then to ascribe to every observation only statistical value? Must we repeat it a great number of times to find out whether we always get the same result and the observed quantity is a characteristic of that state, or, if not, what are its statistics in that state? This is quite apart from "lack of precision" or "observational errors," we are not talking about *them*, we are speaking of an *ideal* observation.

Well, with regard to the state we were presented with, we must. Yet if we abided by this merely statistical meaning of observation, we should

(a) be in too flat contradiction with what observation has always meant and still means to anybody;

(b) we should also be in contradiction with many experiments;

(c) we should *only by good luck* ever hit on a state of which we *know* the wave-function, to wit if, presented with a system in a given state we happen to hit on a measuring device which on frequent repetition with the system in that same state always gives the same answer: then and only then would we know that the wave-function is an eigenfunction of the associated operator.

To explain (a): a measuring device is not some kind of knocking the system about; it is a carefully thought out procedure calculated to give some definite information—*information that we have not got yet*. If we find that, when repeated on the same system in circumstances which exclude or render it unlikely that the quantity we are out for has changed in the meantime, if, I say, in such conditions our device gives a widely different result, grave suspicion is aroused against it. If on *this* kind of repetition it would be found, that *sometimes* the result fluctuates wildly, *sometimes* it is sharply repeated, we should condemn the device as being entirely unsuitable for getting the information we

---

[4] It is essential for the current interpretation that this should be possible. If it is denied the entire interpretation collapses like a house built of playing cards.

are out for. Physics has been built on observations which prove *not* of this type, on *repeatable* observations—and is still built exclusively on such, even though quantum mechanics has become a very fashionable subject in our time.

There is no way out of this impasse but to assume, to declare, to define (whatever you prefer) the following. Observation may (contrary to what it was always believed to do) fail to give us information on the value of the measured quantity *before* the observation was made—the quantity need not have had a definite value. But it definitely informs us of the value *after* the observation. If this were not assumed, we would hardly know *what* information it gives. The wave-function *after* a precise measurement is declared to be an eigenfunction of the associated operator belonging to the eigenvalue that the measurement indicated. This is the only way of getting at a system with a well defined eigenfunction. We shall have to enter on that in detail.

Even this assumption is a great and painful concession of the experimentalist to the quantum theorist, a concession with far-reaching consequences. He thereby gives up half of his confidence in experimental methods. Not the value *before*, but *after* is procured! This includes that in many cases the measuring procedure cannot avoid changing the state of the system profoundly. It includes that, in principle, a measurement is a "Procustation"; it *produces* the value that it enounces as its result, it does not really *find* it in nature. True, it usually does not hit on any value with equal probability; the latter is defined by what we called the "statistics" of this quantity in the original state (which can be found out by repeated measurement on the original state, as described above). The only redeeming feature that saves the *dignity* of observation is, that an *ideal* observation is supposed to yield this *statistics* with the same precision as it repeats the sharp value, when it *is* sharp.

Here I cannot help inserting a remark on the attempts to find in quantum statistics an outlet for free will.[5] If the free decision of anybody were supposed to be allowed to fumble with these statistics, directing it this way or that way, this would mean just as serious an interference with the laws of quantum mechanics as the same thing would have meant in pre-quantum physics with regard e.g. to the Newtonian laws of motion.

---

[5] *Editor's note:* Schrödinger's most extensive discussion about free will and quantum mechanics can be found in: *Science and Humanism; Physics in our times*, Cambridge University Press, 1951.

(Interpreting the Wave Function) • 73

Now let us attend to the details of determining the wave-function by ideal measurements. If there is no degeneracy, that, if there is only *one* eigenfunction to the eigenvalue returned by observation, then this is now the wave-function. Nothing remains undetermined excepted for a phase factor, a multiplying constant of absolute value 1 which is irrelevant for all intents and purposes. What if there are several eigenfunctions to the same eigenvalue?

$$f_1(x) \ldots f_k(x). \quad \underline{\lambda} \tag{7,1}$$

(From now on we definitely regard the one variable x as a representative of all the variables in some representation, configuration space or otherwise). Then, if we do nothing more, we have *not* got a wave-function. Such cases are sometimes dealt with by the method of the "density matrix," which is a generalization of the wave-function that we shall speak of a little later. How are we to proceed in order to get the system into a state in which we know its wave-function? *We must make some further measurement which does not invalidate the outcome of the first.* What kind of measurement? We'll see. Say it has the result μ. Now let us again scan the possible wave-functions to see whether the choice has not become unique. Since the previous result is to be preserved, it must be an eigenfunction to eigenvalue λ of the first and an eigenfunction to eigenvalue μ of the second associated operator. For the first it is necessary and sufficient that it be a linear aggregate of $f_1(x) \ldots f_k(x)$. Thus it is *necessary* to choose the second measurement so that its associated operator has at least one linear aggregate of these $\underline{f}$'s as an eigenfunction. But is this *sufficient*? Certainly not. Remember our μ-measurement is still quite undetermined! If you want it safely to produce a linear aggregate of the $\underline{f}$'s, you must *demand* two things, viz.

(i) that there be k independent linear aggregates of the $\underline{f}$'s which are eigenfunctions of the μ-operator.

(ii) that the μ-measurement must produce one of the eigenvalues belonging to *them*.

As regards (i), we may straightaway take the f_k to mean those linear aggregates, eigenfunctions in common of the two operators. As regards (ii) I do not think it can be consistently *deduced*, because we *have no* wave-function, we are just about to determine it; and all our previous statistical stipulations refer to the wave-function. One ought, I think, at this stage to avoid the attitude: we *do not know* the wave-function, but we know that it must "be" some linear aggregate of the $\underline{f}$'s. The wave-function *is* information of a certain kind. As long as part

74 • Chapter 2. Transformation and Interpretation

of this information is missing the wave-function *is* not this *or* that, but simply *is not*. So we ought to state frankly that we adopt a plausible new axiom, albeit perhaps only provisionally until we enter into the discussion of the density matrix; this is:

If in the course of determining the wave-function we have by previous measurements restricted it to a sub-space (of function-space) spanned by a set of eigenfunctions of a certain operator, the measurement associated with *this* operator does not relax, but, if anything, narrows this restriction.

In our case this narrowing will take place if the μ-values belonging to the k functions (7,1) are not all the same. If the one that we have found has only *one eigenfunction among the* f_k, this is the wave-function. If it has more, we must repeat the procedure, making a third measurement (ν) whose associated operator bears to the remaining set of eigenfunctions in *common* with λ and μ the same relation as the μ-operator had to the f's. And so on, until we are landed with one function only.

The set left over after every measurement consists of eigenfunctions in common to the operators associated with this and all previous measurements. If we wish to prepare a set of observations (as our $\lambda, \mu, \nu, \ldots$) *in advance* that will fulfil the required conditions in any case, irrespective of the numerical results of the single measurements, and is sure to lead to a full determination of the wave-function, we must choose the operators so that together they have one and only one *complete* set of wave-functions in common, which is non-degenerate, as it were, with respect to their conjunction. Such a set of operators is entirely equivalent to one non-degenerate operator and is said to correspond to a complete set of commuting observables; the "commuting" has a physical and an analytic meaning. The first means that for the result it is in a certain sense irrelevant, which measurement is carried out first; the second means that the operators commute. We shall first discuss the analytical meaning.

If a function f is an eigenfunction in common to two operators ϑ and P, they may be said to commute on (or with respect to) this function; for either of them amounts then to multiplying f by a constant, handing it over to the other operator essentially unchanged; and constant factors commute. It follows that if ϑ and P have a *complete* orthonormal set of eigenfunctions in common, they commute on every function of the set and therefore on every developable function—they simply *commute*, like $\frac{d}{dx}$ and $\frac{d^2}{dx^2}$ or x^2 and $y\frac{d}{dy} + x^3$.

(Interpreting the Wave Function) • 75

Inversely, if ϑ and P commute, there is at least one complete orthogonal set of eigenfunctions in common to them. This is proved as follows.

Let f be an eigenfunction of ϑ, eigenvalue λ:

$$\vartheta f = \lambda f$$
$$P\vartheta f = \lambda Pf$$
$$\vartheta Pf = \lambda Pf \quad \text{(since } P\vartheta f = \vartheta Pf \text{ by association)}$$

Thus Pf, unless it vanishes, is an eigenfunction of ϑ with the same eigenvalue λ. The analogous thing holds, of course, with the rôles of P and ϑ exchanged. Now we wish to obtain from f eigenfunctions in common. We develop our f into a series of eigenfunctions of P, say g_k's:

$$f = \sum_k a_k g_k$$

We want this development for the present purpose to be understood in the way that terms which have the same P-eigenvalue are clamped together (we do not bother about normalization now). Let me remind you that even so such a series cannot vanish unless all the single terms vanish (uniqueness of development with respect to eigenfunctions). Operate with P

$$Pf = \sum_k a_k \mu_k g_k \,.$$

This might vanish identically. If so, we have reached our goal, for f itself is then an eigenfunction in common. If not operate with ϑ, remembering that Pf is an eigenfunction of ϑ, eigenvalue λ:

$$\vartheta Pf = \sum_k a_k \mu_k \vartheta g_k = \lambda Pf = \lambda \sum_k a_k \mu_k g_k$$

Hence

$$\sum_k a_k \mu_k (\vartheta g_k - \lambda g_k) = 0$$

Now remember that ϑg_k is an eigenfunction of P of eigenvalue μ_k (by the theorem proved at the outset, only with exchanged rôles). Hence in this sum again all the g's with the same eigenvalue are clamped together. Since it is zero, its single terms must vanish. Hence for every k one of three things must happen:

$$\text{either } a_k = 0, \text{ or } \mu_k = 0, \text{ or } \vartheta g_k = \lambda g_k$$

We have thus proved: these eigenfunctions of P which turn up with a non-vanishing coefficient in the development of f (understood as described above!) are eigenfunctions in common, except perhaps those with eigenvalue $\mu_k = 0$. But it holds also for them. To supply the proof for them, repeat it with P − <u>a</u> in lieu of P, where <u>a</u> is any real non-vanishing constant. (We could have disembarrassed us of the case $\mu_k = 0$ from the outset, replacing P by P − <u>a</u>, where <u>a</u> is a non-eigenvalue of P; but there need not be one!)

Generalizing and summarizing: *all* eigenfunctions of P that occur in the development of *any* eigenfunction of ϑ are eigenfunctions in common. Taken all together, are they *complete*? There may be numerous duplications, and dependences, but in their entirety they must be complete, because the f's are. Any function can be developed with respect to the f's, and therefore, if the f's are replaced by their expressions in the g's, also by the g's *in question*. Hence the latter can be made into one complete system by ruling out duplications or linear dependences, which could only occur between such as have the *same* eigenvalue both for P and for ϑ. This proves the theorem announced at the beginning of this paragraph.

I should like to emphasize a crucial point which is not very often mentioned. It is essential that our functions f and g are true *eigen* functions, not just functions that are reproduced by the operator in question up to a multiplying constant. By this I mean, that they must be continuous and, in the case of differential operators, have the required number of continuous derivatives. If this is not the case, then such a simple thing as a $\frac{d}{dx}$ and $\frac{d^2}{dx^2}$ produces δ-functions. Though this seems to matter only at certain points, *it encroaches with the uniqueness of development* all over the place, since a δ-function has the well-known bilinear development. Of course, in some cases, as we know, the eigenfunctions themselves are δ-functions. We cannot help that. We must in these cases keep them strictly within the prescribed libertinage and not allow any other tomfoolery either to them or to any other function on that account.

To extend the theorem to more than two commuting operators, assume that you have found a complete set of eigenfunctions in common to the commuting operators $\vartheta_1, \vartheta_2, \vartheta_3, \ldots$. We wish to show that from it we can find one that is also shared by one further commuting operator P. We take any one of the first set, call it again f, and proceed exactly as before, viz. we develop it with respect to the g's of P in the same fashion. Then we can show in exactly the same way that any g_k which turns up with a non-vanishing coefficient in any of these developments is also an eigenfunction of ϑ_1; and also of ϑ_2; and so on. It follows that the required set can be produced for three operators, because it can be produced for two; and hence for four, for five and so on. Hence a finite number of commuting operators have at least one set of eigenfunctions in common. If in addition there are no two functions in this set, which for each of the operators belong to the same eigenvalue, the set is unique, because then obviously no linear combination is allowed which would not destroy its property. Then we say the associated physical operations serve to determine a complete set of commuting observables.

I now come to speak of the second, the physical meaning of this epithet "commuting." About this meaning several thoughtless statements are in vogue, the discussion of which shall be illuminating. But I had better first say what it really means. It means what we have in a different context already said earlier, namely the following. If by an experimental measuring device the wave function has been restricted to a certain linear subspace of Hilbert space (to wit the subspace spanned by the eigenfunctions belonging to *that* eigenvalue of the associated operator that resulted from the measurement) then the subsequent carrying out of another measurement whose associated operator commutes analytically with the first cannot relax but, if anything, narrows the confinement. This follows obviously from what we had, in the previous context, laid down as an axiom; for we have shown that the premise of this axiom is fulfilled, namely the eigenfunctions of the first operator that span the subspace in question can be linearly reshuffled so that they are also eigenfunctions of the second which then span the same subspace. It follows that for the purpose of eventually confining the wave-function to just *one* dimension, i.e. determining it, the *order* in which we carry out the complete commuting set of observations is irrelevant. The simplest way of thinking is to use from the outset the unique set of eigenfunctions in common to all of them. Then the inevitable result of carrying out all these observations in any order is one and only one of those functions.

I now come to speak of the thoughtless statements on the same point whose discussion, I said, would be illuminating. They take the form that measurements whose associated operators commute are independent of each other, or do not influence each other or—worst of all—that the result does not depend on the order in which they are carried out.

It is difficult to see what the alleged independence of the measurements or the contention, that they do not influence each other should mean, if not, that the result of the first does not alter our expectation with regard to the second. This is patently in contradiction with our assumptions. To put it drastically by an example: if we are presented with a (non-relativistic) hydrogen atom about which we know nothing, its moment of momentum may turn out to be any one of its eigenvalues ($n(n + 1)h/2\pi$, with $\underline{n}$ an integer). But if, before measuring it, we first measure its energy and find the atom in the k'th quantum state, then we are sure that a subsequent measurement of the moment of momentum cannot give $n > k$. Inversely, the energy may, from the outset, be anything. (any $-Rh/k^2$, where R is the Rydberg-constant). But if we first measure the moment and find $\underline{n}$, then $\underline{k}$ must turn out $\geq n$ on a subsequent measurement. This is a childish example, since nobody has ever measured the energy or momentum of a single hydrogen atom and I daresay nobody ever will. But as an example it will do.

I said worst of all—and, I may add, most illuminating of all—is the assertion, that the *results* do not depend on the order in which the two measurements are carried out. *This might be true in spite of what has just been said.* In the preceding example we always get a couple $\underline{n}$, $\underline{k}$, such that $k \geq n$. No *immediate* contradiction is involved in the belief that in every single case we should have found the same couple if we had carried out the measurements in the reversed order. But how are we to tell, since even the single measurement gives, in general, different results when carried out on the system in the *same state*? So will the two subsequent measurements, even when carried out in the *same* order. There *might* be something at work that predestines both results each individual case, predestines it in such a way that the order of executions does not interfere with it. But we cannot check on this. Our situation is somewhat similar to that of a physician who prescribes a medicine, upon which the patient recovers. It is impossible to decide whether he would have also recovered from this individual attack of disease without taking the medicine.[6]

Yet since this "belief in predestination" is not immediately self-contradictory, it is advisable to follow up its consequences. It means, that all the characteristic values we obtain while carrying out a well-planned complete set of commuting measurements on a system have been characteristics of that individual system in the original "virgin" state in which it had been presented to us. But what holds for one such well-planned set, must hold for any other. And there are many.

---

[6] The physician relies on statistics. *We cannot*: the statistics are against us, anyhow.

For instance in the case of one degree of freedom with which we dealt in earlier sections, x by itself is one, p by itself is one, $x^2 + p^2$ is a third, and there are many, many others. For each of them we should have to admit a "predestined" value (or values) characteristic of the system already in what I called its virgin state. These values are not in simple algebraic relation to each other, for instance x can have any real value, so can p, but $x^2 + p^2$ must be an odd integer. That "belief" therefore would compel us to ascribe to a system simultaneously a host of characteristics, not less but many more than classical physics ascribed to it; they are only not on the surface, they are unravelled only by suitable measurements. And we must not assume—always following that "belief"—that by unravelling as many of them as we simultaneously can, i.e. by executing a complete commutable set of measurements, the others are effaced. For what I called the virgin state was really *any* state. We might now hand our system over to another physicist who investigates it according to a different plan and also arrives at definite results for *his* set of commuting observables. The kind of wave functions arrived at after carrying out such a plan is, of course, entirely dependent on the plan; for instance in the simple "x-case" it is a $\delta(x, x')$ on the x-plan, an e^{ixp} on the p-plan, an $h_n(x)$ on the $x^2 + p^2$-plan. This wave function, if we take together all our previous assumptions and the "belief" would have the following properties

(i) it gives correct information on the probable outcome of any subsequent measurement to be carried out on the system;

(ii) yet it is not characteristic of the actual state of the system, only of our incomplete knowledge thereof, since for many observables (some would say for half of them, some would say for the overwhelming majority) the wave function gives us only a statistics, though, according to the "belief," all have "predestined" values anyhow.

Is it still a more or less irrelevant question of philosophical attitude whether we choose to accept the "belief" or to reject it? No. If our previous assumptions as regards the actual behaviour of nature are adequate, the "belief" is definitely inadequate, it may and must be rejected on physical grounds.

Let us envisage just one particular experimental device—call it "observing λ." If our interpretation of quantum mechanics is adequate, then we are sometimes faced with a state in which this λ-observation gives always the same value, sometimes with an equally well defined state where the λ-observation, repeated on the same state, gives widely different single values, but a definite statistics. We must regard it as equally well defined, because some other observations, say even some complete commuting set of observables have *repeatable* values, just as λ and some other observables had in the first case. The two cases are of entirely equal standing. Only the particular variable λ in which we have selected is among the sharply repeatable variables in one case, among the "merely statistical" variables in the other case. Furthermore the *statistics* of λ is repeatable in any case. If you make 10,000 observations and then again 10,000, they are supposed to give the same statistics, apart from the chance fluctuations (on account of the number not being infinite). *Furthermore*: if we make but one λ-observation and repeat it *on the same system*, the very same appliance now all of a sudden repeats sharply the value of the first observation.

If this is correct, and if the "physical state of a system" is to mean its observable behaviour towards any kind of physical appliances and is deemed to be known only by observing this behaviour and by nothing else, then—if words are to mean anything at all—the *statistics* of λ must be regarded as a well defined characteristic of the first state just as the *value* of λ naturally is in the second state. We must *distinguish* between the two states in which the behaviour is entirely different. To register this difference merely as a difference of our state of knowledge about the system is entirely inadequate for conveying to anybody the difference in behaviour that, according to our assumptions, we have to regard as an actual fact. Undeniably our knowledge is different in the two cases and undeniably this our knowledge, and nothing else, is laid down in the wave function. But this does not mean that nothing else is different. For in both case our knowledge is in accordance with the actual behaviour of the system, which is different. No "metaphysics" is dragged in, if we call this a *real* difference. If I count the money in my purse today and find to have £2.10s., and if I count it tomorrow and find sevenpence-halfpenny, there is a difference in knowledge—but also in facts; nay we only call *knowledge* what we believe to be in accordance with facts! *Thus the wave-function is believed to embody facts*. It is the marvellous tool that is supposed to embody all facts concerning the behaviour of the system, not only the values of those observables that have sharp values, but also the statistics of those that have not. Quantum mechanics must regard it as the full counterpart of the complete classical description of the system.

It is clear that we speak here of the ideal, attempted mental shape of the new theory—which it is necessary to know. The actual performance needs must remain imperfect just as the classical picture always was.

To me the most interesting feature of quantum mechanics is that it suggests, and actually has brought about a *shift of interest* which is, I believe, felt by everybody, but deserves to be mentioned explicitly. The fact that very often "only a statistics" is available is sometimes felt as an opprobrium. In particular this is cumbersome in the case of continuous variables as x or p, which formed the base of the old mechanics, while in the new one they can, to say the truth, never be exactly known: not x, because p would then be more likely than not, to surpass any given finite value; that is to say, after a sharp measurement of x the object has jumped away; not p, because the vice-versa holds now for x, so that the object is infinitely unlikely to be still in the precincts of the laboratory. Correspondingly the eigen-function of a continuous variable, in the frame where this variable is "on the diagonal," is always a δ-function—this most improper function whose use we should fain avoid if we could.

But are we genuinely interested in the precise value of a variable like x or p? The astronomers always said they were and still are. But are they really, except for practical purposes? I think not. The genuine theoretical interest was and is, to confirm the laws of Newton, and recently those of Einstein. We are interested in general laws not in special facts.

Now in this respect quantum mechanics has brought in something entirely new: *There are observables with a discrete spectrum.* To them the interest has shifted. Not only is the mathematical apparatus as apt to deal with discrete variables as it is awkward for handling the continuous. But every special theory of a physical object predicts definite discrete eigenvalues for certain observables. It is extremely interesting and relatively easy to compare them with experiment. This means comparing directly the *nature* of the model of our thoughts with the nature of the physical object, as against comparing *states*; by the latter I mean, generally speaking: we put the model in a definite initial state, investigate how it will "move" according to the theoretical laws, call this the "predicted" motion and observe whether the physical object actually "moves" in this fashion. This was the classical method, that followed the grand example of astronomy in predicting the movements of the celestial bodies. It has greatly lost importance in contemporary physics.

82 • Chapter 2. Transformation and Interpretation

8 • (Laws and Changes with Time)

We must now attend to the way in which the state of a system changes with time. This is, of course, the most interesting and the most relevant part of the theory. And it remains of primary interest whether or no we accept the preceding arguments about the informatory rôle of the wave-function and the way it is determined by measurements or rather "forced into pattern" by measurements. I actually do *not* think this to be the appropriate way of looking at things. I have explained it here a) because the mathematics is necessary anyhow, b) also because it is the current view and, c) because one can only with some clarity say one disagrees with a view, after one has explained it. I have already alluded to some of the reasons of disagreeing.

(i) There are no general rules for the association of linear operators and experimental measuring devices; it is next to impossible to establish such general rules. Unless I am quite sure of the operator, I cannot use my wave-function for prediction even if I have one. But what is worse : the apodictic contention that the measuring device forces the wave-function into a quite definite pattern, requires a minutiously precise description of the kind of action you mean. You need only open a book like e.g. the good old Kohlrausch to find that there is usually more than one method for measuring the same thing, there is often a long list of quite different methods. It is highly improbable that they should all have precisely the same effect on the physical object in question.

(ii) With all the milliards of dollars and pounds the governments now lavish on the promotion of physics, I maintain that nobody has ever yet been able to procure a physical system whose wave-function was known to him or to anybody else. Therefore the whole idea of predicting the results of future observations from the wave-function is, in this dogmatic form, a dream of positivist philosophasters. The wave functions are mental material for building analytical pictures of real objects in one's mind and performing *thought experiments* on them; the results of these thought experiments are obtained by ample use of mathematics and are then compared with the results of real experiments. This is precisely the same pattern physics has always followed, also in the pre-quantum era, when one worked with atoms, molecules, electrons and light waves. The armoury has been enhanced and the analytical methods have changed, that is all.

(iii) We shall now have to study the law according to which the wave function changes as time goes on—the wave equation. If one accepts this law—and it is universally accepted as a general law—one must stick to it. It must not be occasionally infringed upon by a man making a measurement. Hence the changes of the wave-function that are supposed to be brought about by a measurement must be thought of as being controlled by the same law, i.e. the wave equation; not automatically, of course; the measurement must be regarded as a physical interference with the object, an interference that changes the wave operator in such a way that the supposed change of the wave function ensues, as time goes on. Quantum theorists bother very little about accounting for the change in this physical way. When they are interrogated and cornered they eventually take refuge to saying: one must not call it a physical change, it is only a change in our knowledge. I consider this an unfair subterfuge—or plainly: nonsense. As I explained the other day: knowledge is only knowledge in virtue of its agreeing with some reality. If my knowledge changes while the corresponding reality remains the same, it either *was* a mistake or it *becomes* a mistake. There cannot be two different "knowledges" referring to the same reality.

I said quantum physicists bother very little about accounting, according to the accepted law, for the supposed change of the wavefunction by measurement. I know of only one attempt in this direction, to which Dr. Balazs recently redirected my attention. You find it in John von Neumann's well-known book. With great acuity he constructs one analytical example. It does not refer to any actual experiment, it is purely analytical. He indicates in a simple case a supplementary operator which, when added to the internal wave operator, would *with any desired approximation* turn the wave function as time goes on into an eigenfunction of the observable that is measured. He found it necessary to show that such a mechanism is *analytically possible*. The idea has not been taken up and worked out since—in about twenty years or more. Indeed I do not think it would pay. I do not believe any real measuring device is of this kind.

But let us now take up the study of the wave-law. We did mention, on the occasion of interpreting $p = -i\frac{d}{dx}$, that quite in general

$$-i\frac{\partial \psi}{\partial t} = H\psi \tag{8,1}$$

controls the time dependence of the wave function, thus of the state. Here H is a certain Hermitian operator, which in the early applications was always obtained by replacing in the Hamiltonian function every momentum p by $-i\frac{d}{dx}$, or more generally $-i\frac{d}{dq}$, where q is the canonical conjugate to p. This procedure gives the so-called Hamiltonian operator in the "configuration"-frame. Some care is needed in this "replacement" because in the classical H-*function* p and q commute, while the operators do not. One has to take care of two things, viz. a) the H-*operator* is to be associated with the energy of the system, therefore as all our *associated* operators it must be chosen Hermitian or self-adjoint; b) if any "rules for translation" are given, they must be *invariant* to quantum transformations. In the case of a mass-point or a system of n mass-points both requirements are taken care of by giving the rule for the configuration frame in Cartesian coordinates, where it is simple, and by stipulating that for any other frame H is to be transformed as any other observable. In Cartesian coordinates, for slow motions, coordinates and momenta are separated; for one mass point

$$\frac{1}{2m}(p_x^2 + p_y^2 + p_z^2) \to -\frac{1}{2m}\Delta \tag{8,2}$$

$$V(x,y,z) \to V(x,y,z)$$

For rapid motion, there is a complication, because V is usually, at least very often, the electric potential, and it would be inconsistent, not to take into account the vector potential. But with it the classical Hamiltonian is known to be

$$\sqrt{m^2 + (p_x - eA_x)^2 + (p_y - eA_y)^2 + (p_z - eA_z)^2} + eV \tag{8,3}$$

where e is the electric charge. You may, of course, prescribe the replacement in this form and insert the operator in (8,1), giving

$$\left(-eV - i\frac{\partial}{\partial t}\right)\psi = \sqrt{m^2 + (p_x - eA_x)^2 + \cdots}\,\psi$$

But now you have trouble to get rid of the square root, because the A's must in general be regarded as depending on time, so that the two operators, the left and the right, do not commute on ψ: they give, as the equation tells you, the same result when applied to ψ, but not necessarily when applied to the function which results from applying them to ψ. I will not insist on the point here, since the difficulty has been overcome by Dirac's method of rationalizing the square root—which is, of course, not just a mathematical method, but a new and important physical theory. In the following we regard H as independent of time, unless the contrary is stated.

If one uses other than Cartesian coordinates, the simplest thing is, to write Δ in the invariant form, that applies to any space-coordinates, for instance polar-coordinates. This gives one the clue for how to proceed in the case of mechanical constraints, such as the "rotator" or any rigid configuration of masses. In the case the (non-relativistic) kinetic energy has always the form

$$T = \tfrac{1}{2} \sum_i \sum_k g_{ik} \frac{dg_i}{dt} \frac{dg_k}{dt} = \tfrac{1}{2} \sum_i \sum_k g^{ik} p_i p_k,$$

where $P_k = \left(\frac{\partial T}{\partial \dot{q}_k}\right)$ all q's and the other q̇'s constant and the g's are the functions of the q_k only. *The replacement must not be made in this form*, but in the form

$$-\tfrac{1}{2} \sum_i \sum_k g^{-\tfrac{1}{2}} \frac{\partial}{\partial q_i} g^{ik} g^{\tfrac{1}{2}} \frac{\partial}{\partial q_k} \qquad (8,4)$$

which is invariant to coordinate transformations that treat ψ as an invariant and which obviously gives T on replacing $\frac{\partial}{\partial q_k}$ by ip_k ((8,4) is the generalized Laplacian).

We come to discuss the general meaning of equation (8,1). If H does not depend on t, it is easy to write down its formal solution, viz.

$$\psi = e^{itH}\psi_0, \qquad (8,5)$$

where ψ_0 is that function of the configuration coordinates (x say) that ψ is at t = 0, and the exponential means the power series. This shows, as is very often stated, that the change of the wave-function within any given time t is of the same kind as is produced by a certain change of frame—in close analogy with classical mechanics, where this type of change is called a contact transformation. How is this clear from (8,5)? Well it is known and it is easy to show, that the exponential of an Hermitian operator is a unitary operator.[7] A unitary transformation is defined as a rotation in function space—but a rotation generalized to the complex , so that it leaves the "functional absolute square" of

---

[7] The *vice-versa* is *not* true.

any function invariant. An alternative definition of a unitary operator is: its transposed (or adjoint) conjugate complex is its reciprocal. In the latter form you grasp the connection between H being Hermitian and the exponential being unitary at once: H is its own transposed conjugate and e^{-itH} is obviously the reciprocal. To see the invariance of the "square" of ψ, use

$$-i\frac{\partial \psi}{\partial t} = H\psi \quad \text{and} \quad i\frac{\partial \psi^*}{\partial t} = H^*\psi^*$$

$$\frac{d}{dt}\int \psi^*\psi \, dx = \int \left(\frac{\partial \psi^*}{\partial t}\psi + \psi^*\frac{\partial \psi}{\partial t}\right) dx$$

$$= -i\int (\psi H^*\psi^* - \psi^* H\psi) \, dx = 0$$

since H^* is the adjoint.

And this "zero" means identically zero, since the same argument applies at the next moment or at every moment of time ("conservation of normalization").

Still (8,5) looks rather different from the changes of frame we had envisaged in the transformation theory, and we shall only gradually succeed in grasping just how they match. First let me emphasize that there is one momentous difference—a difference of exactly the same kind as between the following two things; after you have referred a body to a Cartesian frame you may change its coordinates *either* by referring it to another frame, *or* by taking the body itself in your hands and shifting and turning it into a new position. The big difference is, that in the second case its relationships to other bodies are changed, in the first case they are not.

In our case, when the wave-function "changes" by a change of frame, we have also to "change" the operator to keep what we called the *information* invariant—all these changes meaning really only changes in the *representation* of the *same* wave-function and operators. The change of the wave-function indicated by (8,1) or, equivalently, by (8,5) is a change actually occurring in time. The *situation* changes and hence also the information must change, if it is to remain correct. We must therefore *not* transform the *operators* accordingly, but keep them what they are.

But now comes a rather interesting aspect. We can at any moment *undo* the actual change of the wave-function "screw it back in time, as it were, to time zero" by a *change of frame*. Since this is now merely a change of frame that is to leave the information what it was (to wit referring to time t, *not* to time zero) all operators have now to be transformed as well. This can be done at any moment of time, it can also be done continuously, leading back always to the same wave-function at time zero. The changing situation is now reflected by the *operators*. The time dependence has been entirely discharged upon *them*. We arrive thus at an entirely new meaning of "an operator associated with an observable." The two meanings have sometimes been confounded, because the same symbols were used without due warning. The new aspect is very interesting and fertile. What is so fascinating about it, is that the operators representing physical variables, are now, as it were, restored to be actual *variables*, that change as time goes on and, by doing so, depict the change of the physical situation. Naturally the initial wave-function (the ψ_0 of equation (8,5)) still plays a relevant rôle. There are however kinds of statement which do not depend on it, general statements that hold whatever the initial wave-function. It is for making these statements in the simplest form that this method is so particularly lucid. The mathematical tool is the operational calculus, which we have *not used* in these lectures hitherto (though we have been handling operators all the time).

But this was an anticipation. We must now first get a better insight into the time development of the wave-function according to (8,1) or (8,5). We must come to feel its kindredness to the change brought about by a change of frame (and not only acknowledge the mathematical identity, which we have proved before).

The difficulty to our understanding is that the original wave-function $\psi_0(x)$ and that at any fixed later time $\psi(x,t)$ *seem* to be referred to the same frame, namely the (eigenvalues of the) configuration variables x; only, we are faced now with a different function. Naturally, if we want to regard the thing as a *transformation*, we must just *not* think that way. We must get ourselves—if only temporarily, for the purpose of understanding—to regard it as the *same* function in a different frame of reference. This is the best achieved by envisaging not just the transformation of one particular ψ_0, but of any or all possible wave-functions. Equation (8,5) is a mapping of function space onto itself. Every $\psi_0(x)$ is associated with one $\psi(x,t)$ and *vice-versa* in a one-one correspondence. Are there now perhaps *invariant couples* or at least such for which the correspondence is particularly simple so that they can serve as sign-posts for grasping the general correspondence? Of course there are—there is even a complete set of such, to wit the eigenfunctions of H. Let us call them $\psi_n(x)$ and their eigenvalues E_n, thus

88 • *Chapter 2. Transformation and Interpretation*

$$H\psi_n(x) = E_n\psi_n(x).\tag{8,6}$$

(Whether they are actually discrete or not is in this case immaterial). Then

$$e^{itH}\psi_n(x) = e^{itE_n}\psi_n(x).\tag{8,7}$$

This change is not only particularly simple, but let us note by the way that *if* the wave-function *is* an eigenfunction of H, the time-change is irrelevant for any statistical information; in other words if and as long as a physical system has a sharp value of the energy, nothing *happens* in it. An this remains true, whether this energy level is degenerate or not. We conclude, that we practically never have to do with a system whose energy is sharp.

But this was only by the way. We must not disregard the exponential. Since we can develop any wave-function as a series with respect to the $\psi_n(x)$, we know what happens to any wave-function albeit only in this serial form. The eigenfunctions of the Hamiltonian operators are the "axes of rotation" in functional space. In a "unitary rotation" there are in general no directions that remain entirely unaltered, but there may be, as here, a complete orthogonal frame of "axes" that are only multiplied by constants of absolute value 1. I said "may be" with regard to the infinite case that we are dealing with. For unitary matrices of finite rank there always are such axes.

When dealing with unitary changes of frame we did not characterize them by their axes for the simple reason that they do not always have any, as we shall see presently. We characterized them by a transformation function depending on two sets of eigenvalues. This is the more general thing that never lets us down. In order to bring the two things—the time transformation and the frame transformation—together as closely as possible, we want to know the transformation function in the present case. Unfortunately it is not very simple. It reads

$$f(x, x') = \sum_n \psi_n(x) e^{+itE_n} \psi_n^*(x')\tag{8,8}$$

For indeed, if you multiply *any* $\psi_0(x')$ by $f(x, x')$ and integrate over x' you get $\psi(x, t)$. For $t = 0$ our f is, of course the δ-function. For any other t it is the time-transformed δ-function, x being the variable, x' the parameter. That is so quite in general, at least for a unitary transformation that leads from one continuous frame to another continuous frame. For a discrete frame you must say for δ-function: confined to one point.

Let us now scan the previously studied frame transformations from the outlook we have gained. Very fundamental was the transformation to p-frame, the frame of momenta, equation (5,2):

$$\varphi(p) = \frac{1}{\sqrt{2\pi}} \int e^{-ixp} \psi(x)\, dx. \tag{8,9}$$

The transformation function was pleasingly simple. What are the "axes"? We know a system of axes, the Hermite functions $h_n(x)$. They get multiplied by i^{-n} respectively. Thus the rotation *in* (not around!) each axis is, as it were, by an imaginary angle, a multiple of $i\pi/2$. Two things are noteworthy. First this *frame*-transformation is very nearly a special case of *time*-transformation of an oscillator with the Hamiltonian $x^2 + p^2$. The eigenvalues (in our units) are $(2n + 1)$, thus the time transformation would for $h_n(x)$ be:

$$e^{(2n+1)it} h_n(x). \tag{8,10}$$

If you disregard the common factor e^{it} (which is irrelevant for all intents and purposes) you get our frame transformation for $t = -\pi/4$, and again and again for $t = k\pi - \pi/4$. (No significance is to be attached to the numerical values considering our simplifying choice of units.)— The second remark is, that the choice of axes for the frame transformation is in this case widely arbitrary, since only four different values of the exponential occur.—One might add a third remark, namely that in a quantum-mechanical oscillator the rôles of the coordinate and of the momentum are periodically exchanged—just as in classical mechanics. But this is only a special case of a much more embracing analogy.

The second transformation we had studied in detail was that by the Hermite functions themselves. We replace $\psi(x)$ by the a_n

$$a_n = \int h_n(x)\psi(x)\, dx, \qquad \psi(x) = \sum_n a_n h_n(x). \tag{8,11}$$

This is a unitary transformation, since

$$\sum_n |a_n|^2 = \int \psi^*(x)\psi(x)\, dx = 1,$$

in a customary normalization. What are its axes? Are there functions which by this transformation are multiplied by a constant only? Obviously not. A function on a discrete set of points cannot be the multiple of a function on a continuum.

It is quite clear that the possibility of transforming every unitary rotation in function space to principal axes is lost, if we so to speak *adjoin* to function space "directions" or "vectors" that are determined not by a continuous sequence but by an enumerable number of components. Yet if function space in the first kind of representation is restricted in the way in which it has to be restricted anyhow for our purpose, the second kind of representation is completely equivalent to the first. Indeed in a continuous representation our wave-functions have to be as a rule not only continuous (let alone piece-wise continuous) but differentiable, and, if a relation like (8,5) is to be meaningful, even analytical. On the other hand we have pointed out at the end of Section 7, that we cannot possibly forego discrete representations since it is from them, as a rule, that we draw the most interesting information.

Our time transformation is doubtless more difficult to follow, mainly because we are faced not with *one* transformation, but with a continuous sequence of transformations, depending on t as a parameter. But in another way it is a little simpler than the most general frame transformation because the nature of the case entails that it always leads to a frame of the same *type*—continuous or discrete. Indeed, equation (8,1) or (8,5) may be contemplated in any frame, also in a discrete one; these relations are invariant, provided that the Hamiltonian operator in (8,1) as well as its exponential in (8,5) are transformed in the same way as the operator associated with any observable. We need not stipulate this as a new rule, since H is regarded as an observable, namely the energy; and its exponential follows suit automatically. If the adopted frame is discrete so that $\psi_0(x)$ and $\psi(x, t)$ are, each of them, turned into an infinite set of numbers or "coefficients," H is turned into an infinite matrix, which we may call "quadratic" because its rows and columns are labelled in the same way; the same holds, of course, for the exponential. The frame of eigenfunctions of H, if they are discrete, offers a particularly simple instance, the time transformation taking then the diagonal form (8,7). *To determine the eigenfunctions of* H *is equivalent to solving the general problem of motion.* For all that I know, no other general method of solving the latter rigorously is available.

But we need not choose the eigenfunctions of H, indeed we may not know them. A very frequent occurrence is, that we actually do not know them, but do know a set which may be relied upon as a *near guess*, because they would be the eigenfunctions if some small term in the Hamiltonian were left out. This opens the way to various methods of approximation some of which aim at getting nearer to true eigenvalues and eigenfunctions, others at determining "transition rates" between those nearly but not entirely stationary states. We do not enter on these methods now.

But I do not wish to follow up this interesting aspect of time-dependent operators. The *time-change* of the wave-function at time t:

$$\psi(x, t) = e^{itH}\psi_0(x)$$

is to be undone by a *frame* transformation; how does this frame transformation affect an operator A? Well, the inner product $(\chi, A\psi)$ has to remain invariant.

$$\chi(x, t) = e^{itH}\chi_0(x)$$

$$\int \chi^*(x, t) A \psi(x, t)\, dx = \int e^{-itH^*} \chi_0(x) A e^{itH} \psi_0(x)\, dx$$

$$= \int \chi_0(x) e^{-itH} A e^{itH} \psi_0(x)\, dx.$$

Obviously

$$A(t) = e^{-itH} A e^{itH}\ [8].$$

Let us differentiate with respect to t.

$$\frac{dA(t)}{dt} = -iH \cdot A(t) + A(t) \cdot iH$$

$$i\frac{dA(t)}{dt} = HA - AH\ [9]$$

A very simple connection. Observables that commute with H do not change. They are called constants of the motion.

One may now show, that these *operators* satisfy formally Hamilton's equation of motion. This is usually done as follows. We shall first deal with a simple example, classical mass-point m, q, p, in one dimension, in a potential field V(q).

---

[8] Methodologie.

[9] This disagrees in sign of i with all that precedes, for equation (8,1) on, but our choice of sign in the equation has been patently arbitrary.

$$H = \frac{1}{2m}p^2 + V(q)$$

$$-i\frac{dq}{dt} = Hq - qH = \frac{1}{2m}(p^2 q - qp^2) \ ; \ pq - qp = -i$$

$$p^2 q = p(qp - i) = (qp - i)p - ip = qp^2 - 2ip$$

$$-i\frac{dq}{dt} = -i\frac{1}{m}p = -i\frac{\partial H}{\partial p}$$

$$-i\frac{dp}{dt} = Hp - pH = Vp - pV$$

If $V = q$, this is i. I wish to prove for $V = q^n$ it is: $i\,n\,q^{n-1}$.

$$q(pq^n - q^n p) = -i\,n\,q^n \quad \text{assume this to be so}$$
$$(pq + i)q^n - q^{n+1}p = -i\,n\,q^n$$
$$pq^{n+1} - q^{n+1}p = -i(n+1)q^n \quad \text{that proves it.}^{10}$$

Thus it is true for any *analytic* $V(q)$.

So we have for our time-dependent operators the equations of motion of old mechanics. We can handle them, integrate them, almost forgetting that they are operators, except for the commutation rules,

$$\frac{dq}{dt} = \frac{\partial H}{\partial p} \qquad \frac{dp}{dt} = -\frac{\partial H}{\partial q}$$

The constancy of H is better proved directly. (Notice analogy with Poisson brackets.)

But take our case

$$\frac{dq}{dt} = \frac{1}{m}p \qquad \frac{dp}{dt} = -\frac{\partial V}{\partial q} = -mg$$

(take a constant field of force: $V = mgq$)

$$\frac{dq}{dt} = \frac{1}{m}p = -gt + C'$$

$$q = C'' + C't - \frac{g}{2}t^2$$

$$= q_0 + v_0 t - \frac{g}{2}t^2$$

---

[10] *Editor's note:* a minor mistake in the calculation has been corrected. The typescript began with the assumption that: $q(pq^n - q^n p) = -i\,n\,q^{n-1}$, and the deduction went on accordingly.

where g is a diagonal matrix, multiple of unity. C' is a constant of integration matrix, q_0 is the matrix, to which q reduces at $t = 0$, and v_0 the one to which p/m reduces at $t = 0$.

Regarding the analogy between Poisson Brackets and Commutator: *In old mechanics:* let K(p, q) be any function of the p, q (we will take one dimension only). Then

$$\frac{dK}{dt} = \frac{\partial K}{\partial q}\frac{dq}{dt} + \frac{\partial K}{\partial p}\frac{dp}{dt} = \frac{\partial K}{\partial q}\frac{\partial H}{\partial p} - \frac{\partial K}{\partial p}\frac{\partial H}{\partial q}$$

(or a sum, if there are more dimensions). The Hamiltonian equations are particular cases for $K \equiv q$ and $K \equiv p$. So $\underline{i}$ · Commutator with H is regarded as the Poisson Bracket *with* H. But there is more to it. If the Poisson Bracket vanishes (in quantum mechanics if the commutator with H vanishes) we are faced with a constant of the motion. But in old mechanics, there is the following famous theorem: let K and M both be constants of the motion; *then their Poisson Bracket* is either a constant or a constant of the motion. Proof:

$$\frac{d}{dt}\left(\frac{\partial K}{\partial q}\frac{\partial M}{\partial p} - \frac{\partial K}{\partial p}\frac{\partial M}{\partial q}\right)$$

$$= \frac{\partial}{\partial q}[(K_q M_p - K_p M_q) \cdot H_p] - \frac{\partial}{\partial p}[(K_q M_p - K_p M_q) \cdot H_q]$$

$$= \frac{\partial}{\partial q}[K_p H_q M_p - K_p M_p H_q] - \frac{\partial}{\partial p}[K_q M_q H_p - K_q M_q H_p] = 0$$

In quantum mechanics a little more is true: both the commutator *and* the anticommutator commute with H. In fact the product does—but it is not Hermitian.

What is interesting are the commutability relations of these integration constants.

$$\frac{1}{m}p = v_0 - gt$$

$$v_0 = \frac{1}{m}p + gt$$

$\frac{1}{m}p$ commutes with itself and with gt; so the velocities at time 0 and at time t commute. But now replace v_0 in the expression for q by: $\frac{1}{m}p + gt$, thus:

$$q = q_0 + \frac{t}{m}p + \frac{g}{2}t^2$$

Now q commutes 1) with itself; 2) with the last term; 3) *not* with p.

Hence *not* with q_0. Commute with q:

$$0 = q_0 q - q q_0 + \frac{t}{m}(-i)$$

$$q_0 q - q q_0 = \frac{it}{m}$$

This is interesting in itself. For though we have not yet spoken here about the Heisenberg uncertainty principle, you know that when two observables do not commute they have certainly not a complete system of eigenfunctions in common. If their commutator is a multiple of unity, they can obviously not have a single eigenfunction in common (because on *no* function can their commutator mean annihilation!) From this one can easily deduce the Heisenberg principle; in our case

$$\Delta q_0 \Delta q \gtrsim \frac{t}{m}$$

We had put $h = 2\pi$. How does this enter?

$$cm^2 \approx \sec g^{-1} \cdot \frac{h}{2\pi}$$

$$[h] = g \, cm^2 \sec^{-1}$$

(You see, by the way, how *much* simpler it is, to put $h = 2\pi$, than to drag it along *and* bother to put a bar through it, in order to get rid of the 2π.)

Thus

$$\Delta q_0 \Delta q \gtrsim \frac{t}{m} \frac{h}{2\pi}$$

They get "increasingly less commutable." E.g. with an electron in ordinary units $m \sim h$; if $\Delta q_0 \sim 0.1$ cm, then after a minute $\Delta q \sim 6000$ cm. The interpretation is a delicate point. All our operators *refer* to $t = 0$.

Another point of interest (though it is nearly the same point) is this

$$H = \frac{1}{2m} p^2 - mgq$$

gives

$$-\frac{i}{m} p + igt$$

$$q_0 = q - \frac{t}{m} p - \frac{g}{2} t^2$$

Notice that this does not commute with H. Nor does v_0. This is an intriguing point! No, they are not "constants of the motion," because they contain the time!

Entirely different would be the treatment of the same problem by wave functions. One would have

$$H\psi = \left(-\frac{1}{2m}\frac{\partial^2}{\partial q^2} + mgq\right)\psi = E\psi$$

as the eigenvalue problem. The solutions are—if you put

$$E - mgq = z$$

$$\psi(q) = z^{\frac{1}{2}}\left\{I_{-1/3}(az^{3/2}) + I_{1/3}(az^{3/2})\right\}$$

$$a = \frac{2}{3g}\sqrt{\frac{2}{m}}, \quad \left(=\frac{4\pi}{3hg}\sqrt{\frac{2}{m}}\right) \quad z = 0$$

This is a particularly amusing eigenfunction (very roughly drawn). It is the wave-mechanical picture of a ball thrown up into the air and coming down again.

$$az^{3/2} = \frac{az}{z^{-\frac{1}{2}}} \quad \lambda = z^{-\frac{1}{2}} \quad 1/\lambda = z^{\frac{1}{2}} = p$$

Clearly $p = gt$, $z = \frac{g}{2}t^2$ hence $p \sim \sqrt{z}$; and, of course, the "density" of the wave function is $\sim 1/p \sim z^{-\frac{1}{2}}$.

$$\text{Approximation:} \quad z^{-\frac{1}{4}}\cos[az^{3/2} - \pi/4]$$

It is not quadratically integrable.

3 • NOTES FOR 1949 SEMINAR

• The Problem of Matter in Quantum Mechanics

During the first quarter of this century, the view established itself ever more firmly that matter consists of protons and electrons. This view remained unshaken by the great regrouping of ideas initiated in 1925–26. At a certain stage, it was necessary to include the neutron in the list of constituent particles or rather to unite it with the proton, the two forming states of the same particle[1]. Today, some might doubt whether the meson is to be included—perhaps on the same basis as the photon—and if so under which denomination of the Greek alphabet. Anyhow, there seems to be no doubt, that matter consists in particles. Here I wish to put forth the opinion, that the fundamental conceptions of quantum mechanics, when regarded from the angle of the epistemological basis of the matter concept, disallows us to regard matter as constituted of particles. It has no direct relation to the particle-concept in quantum mechanics, not indeed to the wave concept, but to the concept of observable. This makes a difference.

From the cat-on-the-powder-barrel[2] problem we have drawn

---

[1] *Editor's note:* This idea was promoted by Heisenberg in 1932: the proton and the neutron were considered by him as two *isospin* states of a single particle called "the nucleon." See W. Heisenberg (1932), Z. Phys. 77, 1; English translation in: D.M. Brink, *Nuclear forces*, Pergamon Press, 1965

[2] *Editor's note:* As A. Fine noticed (*The shaky game,* The University of Chicago Press, 1986, p. 78-84), Einstein was the first to introduce the example of the exploding gunpowder, in a letter to Schrödinger, August 8, 1935. This was meant to illustrate the idea according to which quantum superpositions may be amplified up to the macroscopic scale. In his reply, (August 19, 1935) Schrödinger gives his own illustration of the amplification problem: "(. . .) a Geigercounter prepared with a tiny amount of uranium, so small that in the next hour it is just as probable to expect one atomic decay as none. An amplified relay provides that the first atomic decay shatters a small

a conclusion which may hazily be expressed by saying: there are things—coordinates—states of affairs, with respect to which we must never think our information complete enougth to be expressed by a ψ-function. Undeterminateness in the case of these states of affairs is always a true lack of knowledge of the kind known in classical statistics. The reason is not perhaps that a precise knowledge about these states of affairs is *not* obtainable, but—paradoxically—that it *is* obtainable by direct inspection or observation. We may call them *true observables*. They are closely related to what in quantum mechanics is called an observable. We are led to the view (suggested otherwise) that quantum mechanics, while depriving the particles of individuality, ascribes it to the observables, to the observable states. Matter in the meaning of the philosopher really consists of them—not of the particles.

The way in which particles constitute matter is thus a very strange and novel one. We must investigate it more closely. Particles, having no individuality, constitute pieces of matter that have. They do it by giving rise to observables. What we usually call the building material is of a fundamentally different nature from what is built up of it. In current quantum mechanics, this different nature is expressed by the twofold set of mathematical entities we use: vectors and tensors, or wave functions and operators.

How is this transition from particles to individual, truly observable, pieces of matter. Look at it first in a quite naïve way: how is e.g. the pointer-hand of my instrument—or for that matter my pocket knife—thought to be composed of particles? Well, in this fashion: a number of the latter coalesce to form a more extended gathering of a

---

bottle of prussic acid. This and—cruelly—a cat is also trapped in the steel chamber." A similar description can be found in his 1935 paper ("The present situation in quantum mechanics," in: J. A. Wheeler and W. H. Zurek (eds.) *Quantum theory and measurement*, Princeton University Press, 1983). After 1935, Einstein often mixed his own version of the macroscopic amplification and Schrödinger's. Writing to Schrödinger on December 22, 1950, he described the behaviour of a system "of radioactive atom + Geiger counter + amplifier + charge of gunpowder + cat in a box" (in K. Przibam (ed.), *Letters on wave mechanics*, The philosophical library, 1967, p. 39). As we see from the present text, Schrödinger himself adopted this blended version of the cat's paradox in 1949. Other short notes for Schrödinger's Dublin seminar (May 4, 1949) give a more developped but telegraphic-style description of the "cat-on-the-powder-barrel" example: "Ignition, Geiger-counter for α-rays will set it off. Weak source, small hole, 1 per hour, say. You go for lunch. If you had the whole thing described by a ψ-function, the quantum spread would include etc. etc."

The same short notes contain another very interesting sentence about the measurement problem: "Quantum mechanics stops as soon as anything reaches your senses (that has been said in Schopenhauer long ago)"

build or constitution not copied in the immediate neighbourhood by a number of similar gatherings of the same build. This coalescence gives rise to complicated observables—they *are* matter in the meaning of the philosopher.

Here, two remarks must be inserted, two objections met. My pocket-knife is such a system of observables referring in the quantum mechanical way to a gathering of particles. As a rule I carry it in wastecoat pocket and it is the only one of its kind near. But what if I go a new one and the shop assistant profers a tray from under the counter with two dozens indistinguishable equal ones? Is any one of them not an individual? This objection is trivial. If I am not convinced that a weighing scale would find slight differences in weight, I can open the blade of one of them, thereby distinguishing it from the rest. Or I could take a file and make a mark on it. In fact without anything more, they are in different perspective to me. They are in a word at a sufficient distance so as not to be confounded sensually.

Second, one sometimes speaks of the observables referring to a single elementary particle. Hardly ever, really never in all truth; but one does in mind, for the purpose of thought experiments. Then one takes good care to remove in one's thought similar particles from the neighbourhood. Thus a single elementary particle may form a "gathering" of sufficient distinction to produce observables, thus matter. These thought experiments are precisely what has given rise to the wrong idea that even a single particle is an identifiable piece of matter, and thus that matter is composed of particles. We have here reached the limit where an observable might reside, as it were; on a true "sharp" ψ-function instead of a statistical matrix. But, unlike the assemblee of pocket knives, this case never really arises. So we need not bother about it at the moment.

• The Nature of the Elementary Particles

A week or two ago, I got a paper by G. P. Thomson, a lecture he gave last year to the literary and philosophical society in Manchester on determinism in science. He first made what he calls a "more or less orthodox exposition" then launches two ideas "of a highly speculative character." Let me read you what he says about the second one:

> "There are still difficulties with regard to the elementary theory of particles . . . "

If you think of it, this idea *is not so very "un-orthodox."* It only *appears* to be, from the abbreviated way we think of the now current theories and speak of them. It is true that one *very often does deal* with a single particle or with a couple of particles in collision—isolated from the rest of the world, smoothed out for the purpose—looked upon as furnishing nothing but, say, an inertial frame and a magnetic field. But then we know well enough that this is not correct. For indeed, we know that if our system contains *not* only one particle of a given kind, say electrons, we must not allot a separate wave function to each of them, but we must use a *many-dimensional wave function* depending on the coordinates of all the n electrons. And this is *very* essential. For we know we must use a completely antisymmetric function (or in other cases a symmetric one). Only in this way do we take into account the Pauli exclusion principle, or the fact that the electrons "obey" (as one usually says) Fermi-Dirac statistics—or in other cases Bose-Einstein statistics.

Even this many dimensional treatment shows that the "smoothing out the rest of the world" is a much less harmless simplification in quantum mechanics than it was in classical theory. For in the neighbourhood and elsewhere in the world, there *are* other electrons. If, for a moment, we include them in our system, we have to use a wave function in many more dimensions, one that is antisymmetric in all of them. This shows that to *exclude* them is a much more vital incision in, or interference with the actual state of affairs, a much more momentous and, possibly dangerous simplification, than it would be in classical theory. Moreover, we know that the *many-dimensional treatment is in itself not quite satisfactory*, is not the last word, if for no other reason than because the number n of particles under consideration is not a constant. I should say: *not even the number under consideration*, in order not to create the impression that just only the number of those we have to consider changes. No, there is "creation" and "annihilation," as you know.

Well then, you know that we have this very elegant though rather subtle method or device which is mathematically equivalent to uniting into one the cases $n = 0, 1, 2, 3, \ldots$ in inf. of the many-dimensional treatment. It is in three spare dimensions only, but the wave-functions themselves become q-numbers, non-commuting variables—though *not* themselves "observables."[3] (Second quantization or field quantization).

---

[3] *Editor's note:* On the manuscript, there are three interrogation marks in margin, near the sentence "though not themselves observables."

If you write e.g. Dirac's equation in this understanding, then from the outset *it clearly applies to the sum total of all electrons in the universe.*[4] Pauli's principle is safeguarded by certain anticommutation relations between the wave function and its adjoint. That is to say the *elementary* wave functions (to which the actual q-number-ψ's refer as indicating their probability) are automatically antisymmetric with respect to all the electrons[5] of the universe. And again, the incision of which Sir George speaks seems to be a rather daring act, in principle.[6]

[*Additional paragraphs:* Here I think a certain problem emerges about which one might think. The second quantization formalism does imply the "new statistics"—The same that in the other treatment is implied by antisymmetric or symmetric wave functions. *How?* The ψ-functional depends on the whole ψ-field (if you consider *this* diagonal). It gives you the probability of finding this or that ψ-field in three dimensions! This small ψ-field therefore cannot be said to be symmetric or antisymmetric. What in the sum total of the probability enunciations embodies the statistics?

If you go to the a, a†, b, b† variables, you know: the eigenvalue of aa† + bb† = n.

But in the original variables?

In the electron-case you might say, the question is wrongly put, because ψ is complex, thus not an observable. But what in the case of neutral fields, the electromagnetic for instance. That is real. That is an observable. The ψ-functional of the system gives you the probability of finding this or that field-distribution. How is that statistics or symmetry expressed.

But for us at the moment this is a side-issue.]

Both the many-dimensional and the (more satisfactory) second quantization treatment entail or *embody the so-called new statistics.* And this is of course one of their most momentous features. Moreover you see there is an intimate connection between the fashion in which they take care of it and the examples raised against this "cutting off from the rest of the world." If it were not for having to take care of the Fermi-statistics, we could e.g. *in the many-dimensional treatment* use

---

[4] "Think of Maxwell's equation!"

[5] *Editor's note:* Schrödinger adds the following remark, linked to the word "electron" by an arrow: "Question to be solved." This remark is likely to have prompted him to write the bracketed *additional paragraphs* below.

[6] *Editor's note:* A sentence was later added at this point by Schrödinger: "difficult to analyse but after all equivalent to the many-dimensional treatment."

a wave-function which is just the product of the single wave-functions:

$$\psi_1(x_1) \cdot \psi_2(x_2) \cdot \psi_3(x_3) \cdots \quad {}^7$$

instead of:

$$\begin{vmatrix} \psi_1(x_1) & \psi_1(x_2) & \psi_1(x_3) & \cdots \\ \psi_2(x_1) & \psi_2(x_2) & \psi_2(x_3) & \cdots \\ \cdots & \cdots & \cdots & \cdots \end{vmatrix}$$

as we usually have to do. The "smoothing out" by "integrating away" over the coordinates of the particles *not* under consideration would then be a simple affair, just about as simple as in classics.

Perhaps I ought here to insert a remark. I have been speaking of the *trouble* this "cutting away from the rest of the world" causes. This is a *conceptual trouble*, a *philosophical trouble*, not a mathematical one. The word *trouble* has in quantum mechanics for late assumed the meaning that *by apparently sound analytical procedure you produce nonsensical results*. This we do not mean here. The mathematics is simple enough, e.g. you just form the antisymmetric wave-function of as many electrons only as you wish to consider. It is the conceptual justification that is doubtful, the proviso under which it may be legitimate.

The difficulty, as I said just before, is connected with the *"new statistics."* And the new statistics *gives the answer to our question about the proviso*, so I believe. The answer is not such as to be immediately minted into mathematical formulae. That is why it plays a subordinate role in the considerations of quantum physicists. But I believe it is to play a great role in our future attitude towards atomism— what is usually called the particle aspect. If you think about those new statistics rationally and soundly, you see that they are *illogical, when applied to the individuals or to individual events*. Take e.g. this: you play heads or tails every 10 seconds. Pick out one minute. It is illogical to admit that it should be equally probable that there should be either 0, 1, 2, ..., 6 "heads" among these 6 casts (Einstein-Bose statistics)[8]. To make this true, you would have to introduce a most intricate dependance between the probabilities of the single casts. It is equally

---

[7] Works?

[8] *Editor's note:* The problem amounts to looking for the number of distinct ways one can put 6 individuals into 2 states. Here, the individuals are replaced by individual casts and the 2 states are replaced by "head" or "tail." Let $Z = 2$ be the number of states and $N = 6$ be the number of individuals. According to Maxwell-Bolzmann statistics, there are $Z^N = 2^6 = 64$ distinct distributions. This is the *total* number of

illogical to assume that heads can only come once or not at all (Fermi-Dirac statistics)[9]. One might say: envisage the single cast; *there* Fermi-Dirac statistics is just the thing that applies. That is true. But then, the things that correspond to "electrons" in these statistics are obviously not the casts, but the notion "heads." "Heads" is *not an individual* but a property of an individual, namely of a cast. The casts correspond to the possible *states* of the electron. A natural example for Einstein-Bose statistics are the amounts you pay annually the *Mount Street Club* if you have subscribed to their petition. But again, *the shillings and pennies on your bank account are not individuals*. I shall give other and better examples later. For the moment, I wish to interrupt these considerations.

Philosophical considerations about quantum mechanics have gone out of fashion. There is a widespread belief that they have become gratuitous, that everything is all right in this respect for we have been given the marvellously soothing word of *complementarity*, that it is only the detailed mathematical or physical theory which is still at fault. I cannot share this view. In the 20 years of its existence, serious objections have again and again been raised against the current interpretation. *Some of them have not been solved but shelved.*[10] No lesser person than Einstein still withholds his assent. In a letter to Max

---

distinct distribution. The greater relative frequency corresponds to $N_1 = 3$ tails and $N_2 = 3$ heads, for the *partial* number of distinct distributions of individuals reaches its maximum in this case:

$$\frac{N!}{N_1! N_2!} = \frac{6!}{3! 3!} = 20$$

According to Einstein-Bose statistics, the *total* number of (purely quantitative) distributions of non-individual particles over two states is:

$$\frac{(Z+N-1)!}{(Z-1)! N!} = \frac{7!}{1! 6!} = 7$$

and each (purely quantitative) distribution can only occur once. These distributions are thus to be considered equally probable.

[9] *Editor's note:* The analogy is here the same as in the case of Einstein-Bose statistics: casts for particles and "head" or "tail" for states no. 1 and no. 2. *The Pauli exclusion principle*, if applied to casts, thus implies that heads (as well as tails) can only come once or not at all.

In the following sentences, Schrödinger reverts the analogy: "heads" correspond to *non-individual* electrons and casts to its individual states. See E. Schrödinger, "What is an elementary particle?" Endeavour **9**, 109-116, 1950.

[10] *Editor's note:* This expression was also used in Schrödinger's short notes for Dublin seminar (May 4, 1949): "In order to justify myself for going into such outmoded

Born, he formulated his opinion in one marvellously poised sentence: "Of this I am firmly convinced that we shall eventually land at a theory in which the things that are linked by laws are not probabilities but imaged facts, as was taken for granted until lately."

I would like to tell you now about one of these shelved cruxes.

- ## The Difficulties in Interpreting the Blur (Quantum Uncertainty)

    There is a real difficulty in interpreting the uncertainty or the statistical character of the information to be extracted from the wave function. A difficulty which cannot be explained away by "positivist philosophy."

    Say you have a system on which you have carried out a complete set of observables, so that the system is now described by a ψ-function. Envisage an observable Q (*not* of this set, otherwise it would be sharp). The information you get for it from the wave function is a certain blurred distribution (see figure below).[11]

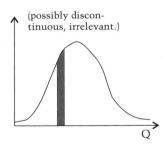

Towards this, there have been two attitudes:

(A) The observable Q has a definite value, but you do not know it, and the ordinates mean the probability of this or that value. Or, if you like, the frequency (prepare the system the same way over and over again, measure Q etc.). *Lack of knowledge.*

---

things as questions of interpretation, I wanted first to show you that there are still objections that have been raised long ago, and not solved but shelved."

[11] *Editor's note:* We find a similar curve in the short notes for Dublin seminar (May 4, 1949). Schrödinger's commentary was the following: "We are speaking of difficulties one encounter in interpreting the 'quantum spread.' The question is just: what does the ordinate mean? (. . .) Not paradox, not antinomia, *aporia* ($\pi\epsilon\rho\alpha$) = no passage; cannot go across, impass."

(B) In this state, Q is genuinely unsharp. It is not a lack of knowledge, it is the nature of the system to have no sharp value of Q—until you measure it.

The attitude now usually adopted is (B), at any rate *not* (A). The latter is called metaphysical. "Thou shalt not and thou canst not bother about what really is." There is no meaning in the question: what *is* the value of Q? The ordinate indicates the probability (or frequency) of *finding* this or that value. That is all that can interest us (Positivist attitude).

Unfortunately, this leads us into difficulties too. *Very unfortunately*, because the fact that it leads into trouble as well does not make the attitude of "incomplete knowledge" more acceptable.

The trouble is this. Let the system be a mass-point. Suppose you have located it at K with great accuracy (coordinates x). It is thus represented by a narrow wave-packet with a great (widespread) blur of wave-numbers or momentum (p) or velocity (v). The positivist attitude is that it is meaningless to say it *has* a definite velocity (unless we measure it).

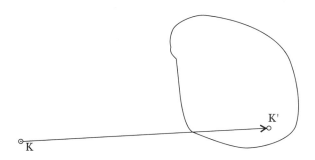

Now envisage the mass-point after a second. Owing to the spread in wave-numbers the "wave parcel" will have enlarged considerably, it has become diffuse. The smaller you had made (!) it at K, the more diffuse has it become now.[12] *Now, measure the place again with fairly great accuracy* (K').

Then, it seems, you have to ascribe *post factum* to the mass point when it was at K a velocity given by an arrow pointing from some point in K to some point in K', thus between very narrow lim-

---

[12] This wave parcel, according to the accepted view, now contains all the information.

its, as narrow as you choose. Not perhaps with certainty, but with an exceedingly high degree of probability.

This is distasteful. *It contradicts the uncertainty principle.*[13] What are the alternatives? You can either admit or deny that the mass-point has come from K to K' "by itself."

(i) If you admit it, then (to avoid the violation of the uncertainty principle), you must deny that velocity is the thing that determine the movement of a body left to itself. That puts you to a loss to tell *what* velocity means at all!

(ii) But you may deny that it has come to K' all by itself. Your second measurement has brought it there.

This *is* of course consistent. Consistent because you now take the same attitude as regards location as you took in K with respect to velocity. But it it then quite clear that a measurement of x affects not only (as is always said) p, but also x itself.

You have not found *a particle at K', you have* produced *one there!*

You must *not* believe (unless you wish to commit the "metaphysical crime") that the particle is launched from K towards K' rather than to any other part of the "cloud." Before the second measurement, it is ubiquitous in the cloud (it is not a particle at all).

By choosing the x-accuracies at K and at K' prudently, you can get still funnier things. Make K *very* accurate, so that the "cloud" becomes an expanding spherical wave. Make K' not too accurate so that it does not blur the velocity *after* the measurement. Then you get at K' a wave-group indicating actually a velocity KK'.

Then your measurement has produced a particle at K' with a velocity *as if* it had been launched some time earlier from K towards K'.

Take it all together: your particle *was* at K. It has *not* been launched in a definite direction, you must not admit that; it would be metaphysics, it would contradict the Heisenberg principle. Yet you *find* a particle at K'—the suggestion being that you produce it there. Anyhow you find it there (if you take care to get some information about its velocity) with a velocity as if it had been launched from K some time earlier towards K'.

---

[13] It is contrary to the view (B) that the point has no definite velocity at K.

The case is not artificially constructed. It happens every day of the week with a β-particle ejected from the nucleus and observed in a Wilson chamber.[14] (At least according to the usual idea which is that you do observe a particle launched from a definite nucleus. This, of course, *is* metaphysics).

---

[14] In fact, this explains why the β-spectrum cannot be sharp!

4 · NOTES FOR 1955 SEMINAR

- **Introduction**

  As many of you know, I disagree with the general attitude now followed in quantum mechanics. I have written up my objections several times, and I will not bore you by repeating it all.

  One central point where I find the attitude inconsistent is characterized by the continued use of the expression "stationary state." It stems from the older form of quantum theory . . . certain of the mechanically possible motions being selected as the only admissible "quantum orbits." But it has turned out, that they correspond to the proper modes of certain vibrations—whatever they may be. This, to my mind, has cleared up the discrete selection, that was so mysterious in Bohr's theory. Moreover the equations controlling the vibrations are linear. Hence their solutions, in particular their proper modes, can be superposed. And the principle of superposition is regarded as a fundamental feature. Then I don't know why the pure proper modes should be any more "stationary" than any superposition, why the system should have any preference for remaining in such a state rather than . . .

  Still worse is the application to statistics. Faced with a system consisting of a great many parts (gas-molecules), people make statistics in assuming that each of these minute parts be always in one or the other of *its* stationary states, let me say it is "on one of its energy levels." Not only, as I said before, is there no reason for assuming this, but it is in violation of that precious principle that the same school of physicists is so anxious to put across, namely that we must never admit anything to *be*, except what we have measured. It would be ridiculous to *think* that we measure all the private energies of the molecules of a gas, for we are—according to this principle—disallowed to ascribe them definite values.

I'll come back to this later. Today I wish to discuss at some length certain considerations concerning the Uncertainty Principle. I wish to make it clear in advance that my aim is *not* (as it might seem at first) to defeat or refute the Uncertainty Principle, rather the opposite. As you know, it is a truism in wave mechanics, and I am firmly convinced it has come to stay—though it may change its name some time.

- **On Measuring Velocities by the Police or Race-Course Method**

In the early days of quantum mechanics, about 1927, the following objection was raised against the Uncertainty principle. Let a free particle be located by measurement at A, and again at a distance a at B, τ seconds later.

Then obviously $\frac{a}{\tau}$ is the velocity with which it has travelled from A to B, thus the one that it *had* at A. The quotient gives this velocity the more accurately it precises the two locations. The accuracy is not impaired, nay it is increased by a very accurate location at A, hence for the initial moment the product of the two Heisenberg uncertainties can be made arbitrarily small, smaller than $\frac{h}{2\pi m}$.

To this Heisenberg answered "yes, but this belated information is of no physical significance; it was not forthcoming at the initial moment at A, could not be used for predicting the orbit; it is only vouchsafed after the orbit is known; and of course it tells me nothing about what happens 'after B'; a very precise location at B disturbs the mobile, its further orbit is not at all determined by its direction $\overrightarrow{AB}$ or by the quotient $\frac{a}{\tau}$—and, of course, for a simultaneous measurement of velocity and ubiety *at B*, my Uncertainty Principle holds."

To this, one would have to say that it is all right, but if one accepts it, one grants to Einstein that quantum mechanical description is incomplete. If it is possible to obtain simultaneous accurate values of location and velocity, albeit belatedly, then a description that does not allow one to express them is deficient. After all (whatever may happen "*after B*") the thing has obviously moved from A to B with velocity $\frac{a}{\tau}$; it has not been interfered with between A and B. So it must have had this velocity at A. And this is beyond the power of quantum mechanical (or wave mechanical) description.

But we can even tighten the argument by devising a more careful measurement at B—not just making it a very very accurate location. We can carefully devise a simultaneous velocity-ubiety measurement at B, in full agreement with the Uncertainty Principle at B, of course, but such that the velocity thus measured at B:

(i) agrees with $\frac{a}{\tau}$ within limits *and*
(ii) determines the future behaviour (after B) within limits, these limits being too narrow for our attributing this velocity to the mobile at A from the point of view of the Uncertainty Principle.

The second point is quite obvious. We can if we like allow a much larger margin of location at B than we did at A, say:

$$\Delta x_B > \Delta x_A \quad \text{(or even } \gg \text{)}$$

Then a Δp_B will be allowed much narrower than Δx_A would allow it, and this p_B certainly controls the future behaviour (after the B-measurement). But one might have the following qualms:

(α) By allowing a comparatively large Δx_B, one has decreased the accuracy of *a*, hence of the quotient $\frac{a}{\tau}$.
(β) Has the measured p_B anything to do with this quotient? Or at least, may it not deviate from this quotient an unknown way, these deviations constituting an uncertainty compatible with Δx_A?

As regards (α), the corresponding uncertainty of the quotient is

$$\frac{\Delta x_B}{\tau}$$

In order to avoid a clash with the Uncertainty Principle at A, one would have to have:

$$\frac{\Delta x_B}{\tau} \Delta x_A > \frac{h}{2\pi m}$$

To fix the idea, take $\Delta x_B = 20 \Delta x_A$. The left hand side is then:

$$\frac{20(\Delta x_A)^2}{\tau}$$

which is arbitrarily small if Δx_A is sufficiently small or, alternatively, τ sufficiently large. So this qualm is abated.

As regards (β), the story is not so simple. *One must use the wave picture* (which is fully admitted by the Heisenberg group, provided one interprets it in their way). Momentum is determined from the wave-length or wave-number. To determine this you must have at least a train of a few (5 or 10 or 20) waves; that is *the Uncertainty Principle!* But in our measurement of p_B there is still another snag. The wave starts from A as a spherical wave, but a very inhomogeneous one and therefore, on account of dispersion, not just a shell, but filling the whole place, the different wavelengh being gradually separated by dispersion.

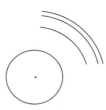

At B the separation is not yet completed, we have an inhomogeneous wave *and this constitutes a further independent reason for the uncertainty of* p_B. We must estimate it. It all depends onto what extent the different wavelengths have already been separated at B.

The wave are separated on their journey with groups velocity, which is the particle velocity; now: $p = mv = \frac{h}{\lambda}$, so

$$v = \frac{h}{m\lambda}$$

The transition time for distance a is:

$$\frac{a}{v} = \frac{am\lambda}{h}$$

During this time two wavelengths differing by $\Delta\lambda$, hence having:

$$\Delta v = \frac{-h}{m\lambda^2}\Delta\lambda$$

will have been separated by a distance:

$$\frac{a}{v}\Delta v = \frac{am\lambda}{h}\frac{h}{m\lambda^2}\Delta\lambda = a\frac{\Delta\lambda}{\lambda}$$

To fix the ideas let us say we use a train of k (5 or 10 or 20) waves for determining λ and therefore p. Then:

$$a\frac{\Delta\lambda}{\lambda} = k\lambda$$

determines the uncertainty $\Delta\lambda$ caused by their lack in homogeneity, thus:

$$\Delta\lambda = \frac{k\lambda^2}{a}$$

But:

$$p = \frac{h}{\lambda}; \quad \Delta p = \frac{-h\Delta\lambda}{\lambda^2}$$

Thus:

$$\boxed{\Delta p = \frac{kh}{a}}$$

So this is the uncertainty caused by the inhomogeneity. *To satisfy us*, it must be smaller than the ordinary one, caused by the finite extension of the wave-parcel, which we have assumed to be $k\lambda$. Calling this *ordinary* uncertainty $\Delta'p$, we have:

$$\Delta'p = \frac{h}{2\pi k\lambda}$$

and we want $\Delta p < \Delta'p$, or:

$$\frac{kh}{a} < \frac{h}{2\pi k\lambda}$$

or:

$$\boxed{\lambda < \frac{a}{2\pi k^2}}$$

You see that this constitutes a certain demand, but not a very exigent demand. Take e.g. $k = 20$, then the distance a must be at least $400.2\pi \approx$ 2500 wavelengths.

So, what is our finding? We have spotted a particle at A with great precision. We find it again at B, at a distance a after time τ, *as if* it had travelled from A to B with velocity $\frac{a}{\tau}$. To make sure we make also an orthodox measurement of velocity at B, we find it within limits equal to $\frac{a}{\tau}$ and the particle will continue its journey, within limits, with the same velocity—and these limits are such as to be irreconciliable with the accuracy of the location at A.

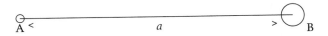

Einstein was inclined to infer from these or similar considerations that quantum mechanical description is incomplete. I am inclined to avoid the incongruity by what I think is the only alternative, viz. there is no meaning in saying that I have observed at B the *same* particle. After an emission process, there is a probability of spotting *a particle* at B after time τ or anywhere else at any other time. This probability (of what Margenau calls a firefly-event) is controlled by the wave-function, which also tells me that if it occurs after τ at B it "has the velocity" $\frac{a}{\tau}$—all this "within limits."

- **Introduction to Statistics**

    My aim is to show that Statistical Thermodynamics can be erected on the basis of the quantum-mechanical level scheme, the scheme of eigenstates or proper vibrations. There is no need for the convenient but absurd assumption that any quantum mechanical system—whether the whole system or its microscopic constituents—should always be in one of its eigenstates. For deducing thermodynamics from statistical mechanics, some fundamental assumptions must be made:

    first an embryonic dictionary *what corresponds to what*.

    secondly a general assumption on the behaviour (what in the classical treatement was "molecular disorder" or "the ergodic hypotheses").

    After that, the dictionary must be extended and it must be shown that everything is consistent and leads to the general (or special) laws of thermodynamics.

    Basic assumption no. 1: E_r, m_r; $S = k \lg m_r$.

    Basic assumption no. 2: on small perpetual perturbations, the square of the amplitudes of degenerate states are on the average equal (this is invariant to quantum mechanical transformations, since it means there is on the average "spherical symmetry" in the subspace of Hilbert-space that belongs to the degenerate eigenvalue—and what is free is just: any rotation in any of these subspaces).

I wish to show in the first place that these assumptions lead to the canonical distribution of excitation strength (average amplitude squares) for any system immersed in a huge heat-bath of given temperature, where the system+heat-bath together are at (or near) a given energy level.

• Planck-Black-Body-Radiation (Without Discontinuity!)

Let us illustrate[1] the general deduction by applying it to blackbody radiation in a cavity with reflecting walls and with a little hole of communication. What are the eigenstates of the whole system, what their energy-levels, what their multiplicities? Let v_k be the frequencies of the normal modes. The levels for v_k are:

$0, hv_k, 2hv_k, \ldots, n_k hv_k, \ldots$ (we'll not now discuss the $n + \frac{1}{2}$ factor).

Any particular eigenstate has, now, the level:

$$E_r = \sum_k n_k h v_k \qquad (1)$$

That is simple enough. But what is its multiplicity? It results from the fact that very many sets of (n_k) give the very same sum E_r. For:

(a) large groups of v_k are practically coincident, so that only the sum of their n_k's is relevant for E_r.

(b) even apart from this you may obviously lower some n_k's and increase others, leaving E_r unchanged.

The first kind of change does not alter the observable spectral distribution, the second does. Since we are interested in the total "intensity" of any particular spectral distribution, we need the multiplicity due *to the first circumstances only*.

---

[1] *Editor's note:* This chapter can be considered as a very late answer to one of the first criticisms Bohr made against pure wave mechanics. During his celebrated discussion with Schrödinger in Copenhagen in october 1926, Bohr "would confront Schrödinger with the fact that without the discontinuous transitions, it was not even possible to maintain Einstein's derivation of Planck's radiation formula" (in: N. Bohr, *Collected Works*, general ed. E. Rüdinger, Vol. 6; *Foundations of physics I (1926-1932)*, J. Kalckar ed., North-Holland, 1985, p. 10). Another (less detailed) version of the same argument can be found in: E. Schrödinger, *Statistical Thermodynamics*, Cambridge University Press, 1962. It was added between the first edition of the book (in 1944), and the second edition (in 1952).

From a well-known asymptotic formula due to Weyl-Debye, there are (in any case, viz. light):

$$\frac{8\pi U}{c^3}\nu^2 d\nu \quad \left(\frac{4\pi U}{3\lambda^3} \text{ scalar}\right)$$

normal modes between $\nu, \nu + d\nu$ (U being the volume); call that:

$$z_k = \frac{8\pi U}{c^3}\nu_k^2 d\nu_k \tag{2}$$

We take them all to be equal and mean by n_k henceforth the sum of these occupation numbers. So we have to distribute n_k over z_k places. From an elementary formula this can be done in:

$$\binom{n_k + z_k - 1}{n_k} \tag{3}$$

ways. Hence the total degeneracy (multiplicity) for given all n_k's is:

$$\prod_k \binom{n_k + z_k - 1}{n_k}$$

and the intensity of this eigenstate is (from our general formula) proportional to:

$$\exp\left(-\frac{E_r}{kT}\right)\prod_k \binom{n_k + z_k - 1}{n_k}$$

where E_r and z_k are given by (1) and (2) respectively. It has a sharp maximum for a particular set of n_k's, and this is what really interests us. We envisage its logarithm:

$$-\frac{E_r}{kT} + \sum_k [(n_k + z_k - 1)\log(n_k + z_k - 1) - n_k \log n_k$$

$$- (z_k - 1)\log(z_k - 1)]$$

and vary it with respect to the n_k:

$$-\frac{\delta E_r}{kT} + \delta n_k \log(n_k + z_k - 1) - \delta n_k \log n_k = 0$$

But,

$$-\frac{\delta E_r}{kT} = -\frac{h\nu_k}{kT}\delta n_k$$

thus:

$$\frac{n_k + z_k - 1}{n_k} = \exp\left(\frac{h\nu_k}{kT}\right) \qquad \text{(pardon the } k!)$$

$$z_k - 1 = \left[\exp\left(\frac{h\nu_k}{kT}\right) - 1\right] n_k$$

$$n_k = \frac{z_k - 1}{\exp\left(\frac{h\nu_k}{kT}\right) - 1}$$

Which is practically Planck's formula.

Is this a new derivation of that famous formula? Not at all, only a new way of looking at a well-known one, due to Bose. He looks at it that way: distributing—within every spectral group labelled by the letter k—n_k photons over z_k oscillators. And there he used and had to use a statistics which would be quite non-sensical if the photons were individual particles, a statistics that we now call Bose-Einstein-statistics.

But what is worse is that on this photon-view one implicitly admits that not only the whole body of radiation but every simple "oscillator" (= proper mode) is always in a state of sharp energy. We have assumed nothing of the kind. The concept of eigenstates and of their degeneracy (or multiplicity) is given and unavoidable in wave-mechanics. It takes the role of the permutation number in Boltzmann's original reasoning.

Notice further that it is the so-called second or *field* quantization that we have used. Its naturally discrete quantum states are at the basis of the "particle picture."

The Fermi Case

It is quite illuminating to try what result one gets by taking the oscillators to be "Fermi-oscillators" with levels 0, $h\nu$ only. Well then the *single* n_k's can only be zero or 1. The n_k referring to the groups of z_k "nearly monochromatic proper modes" can, of course be anything $\leq z_k$ and is distributed over the z_k oscillators in that way: that for n_k of them the upper level, for $z_k - n_k$ the lower, the zero level is taken. This can be done in $\binom{z_k}{n_k}$ ways. So, the total degeneracy is $\prod_k \binom{z_k}{n_k}$ and the excitation strength of that total level that corresponds to *given* n_k is according to our general formula:

$$\exp\left(-\frac{E_r}{kT}\right)\prod_k \binom{z_k}{n_k} = \exp\left(-\frac{E_r}{kT}\right)\prod_k \frac{z_k!}{n_k!\,(z_k-n_k)!}$$

with again: $E_r = \sum_k n_k h\nu_k$

Proceeding as before we determine the maximum of the logarithm:

$$-\frac{E_r}{kT} + \sum_k z_k(\log z_k - 1) - n_k(\log n_k - 1) - (z_k - n_k)[\log(z_k - n_k) - 1]$$

Thus:

$$0 = -\frac{h\nu_k}{kT}\delta n_k - \log n_k\,\delta n_k + \log(z_k - n_k)\,\delta n_k$$

$$\log \frac{z_k - n_k}{n_k} = \frac{h\nu_k}{kT}$$

$$\frac{z_k}{n_k} = 1 + \exp\left(\frac{h\nu_k}{kT}\right)$$

$$\boxed{n_k = \frac{z_k}{\exp\left(\frac{h\nu_k}{kT}\right) + 1}}$$

We have the well-known change from -1 to $+1$ in the Fermi case. Still you know the last formula does *not* apply to a Fermi gas; something is missing. Just the same, the first formula does not apply to the Bose-Einstein gas in general, only to the particular degenerate case of electromagnetic radiation (rest mass = 0). How would we obtain the missing something? At what place must our reasoning be modified?

Let us stay with the Fermi case (the other one is quite analogous). Up to the intensity formula:

$$\exp\left(-\frac{E_r}{kT}\right)\prod_k\binom{z_k}{n_k} \quad \text{with } E_r = \sum_k n_k h\nu_k$$

there is no change possible. But now we have to assume that we can distinguish experimentally the cases of different

$$n = \sum_k n_k$$

and are therefore interested in the sharp maximum for *given* $\sum_k n_k$. This yields at once the missing multiplier—not a constant, but a function of n, T. The customary attitude is of course that $\sum_k n_k$ is the total number of atoms (or atoms) present. It is not so easy to account from our point of view for the fact that *with non-vanishing rest-mass we have to* regard the $\sum_k n_k$ as a known, given quantity that cannot be changed by exchanges with the heat-bath.

But perhaps the ordinary attitude has a similar difficulty to explain why in the black-radiation case the total number of photon is *not* a given number—if one believes in individual photons.

No thermodynamic equilibrium is absolute, may be with the only exception of black radiation.

I should still like to test the formula for the entropy. Let us take the case of black radiation:

$$S = k \log \prod_\ell \binom{n_\ell + z_\ell - 1}{n_\ell}$$

$$\frac{S}{k} = \sum_\ell (n_\ell + z_\ell - 1) \log(n_\ell + z_\ell - 1) - n_\ell \log n_\ell - (z_\ell - 1) \log(z_\ell - 1)$$

we take:

$$n_\ell = \frac{z_\ell - 1}{\varepsilon - 1} \quad \text{with } \varepsilon = \exp\left(\frac{h\nu}{kT}\right)$$

a simple but tedious computation gives you:

$$\frac{S}{k} = \sum_\ell \frac{z_\ell - 1}{\varepsilon - 1} \{\varepsilon \log \varepsilon - (\varepsilon - 1) \log(\varepsilon - 1)\}$$

Is that correct?

Neglecting again in $z_\ell - 1$ the -1, we have for the single oscillator:

$$\frac{S}{k} = \frac{1}{\varepsilon - 1} \{\varepsilon \log \varepsilon - (\varepsilon - 1) \log(\varepsilon - 1)\} \qquad \kappa = \frac{h\nu}{\varepsilon - 1}$$

The test of an entropy expression is always that $\frac{dS}{d\kappa} = \frac{1}{T}$ and $S \to 0$ for $T \to 0$. So express it by κ:

$$\varepsilon - 1 = \frac{h\nu}{\kappa} \qquad \varepsilon = \frac{\kappa + h\nu}{\kappa},$$

thus:

$$\frac{S}{k} = \frac{\kappa}{h\nu} \left\{ \frac{\kappa + h\nu}{\kappa} \log \frac{\kappa + h\nu}{\kappa} - \frac{h\nu}{\kappa} \log \frac{h\nu}{\kappa} \right\}$$

$$\frac{S}{k} = \frac{1}{h\nu} \left\{ (\kappa + h\nu) \log(\kappa + h\nu) - (\kappa + h\nu) \log \kappa - h\nu \log h\nu + h\nu \log \kappa \right\}$$

$$\frac{S}{k} = \frac{1}{h\nu} \left\{ (\kappa + h\nu) \log(\kappa + h\nu) - \kappa \log \kappa - h\nu \log h\nu \right\}$$

For T → 0, κ → 0 and our S → 0. Now differentiate:

$$\frac{h\nu}{k} S = (\kappa + h\nu) \log(\kappa + h\nu) - \kappa \log \kappa - h\nu \log h\nu$$

$$\frac{h\nu}{k} dS = d\kappa \log \frac{\kappa + h\nu}{\kappa} = d\kappa \log \varepsilon = d\kappa \frac{h\nu}{kT}$$

So, this is quite all right.

5 · FROM A LETTER TO ARTHUR S. EDDINGTON, MARCH 22ND 1940

What happens with me,[1] is that I am not so deeply impressed by the alleged grand philosophical revelations which especially quantum mechanics is supposed to have brought us. I am just about 10–15 years older than most of the enthusiastic champions of that new positivist outlook—and I was born and educated in Vienna with E. Mach's teaching and personality still pervading the atmosphere. I was devoted to his writings, which I read practically all before I could know a word of the 1913 Bohr theory—maybe about the same time when we were initiated into the restricted theory of relativity. Just as strong or even stronger than Mach's was in this time in Vienna the after-effect of the great Boltzmann, whose splendid pupil and admirer Hasenöhrl had just taken Boltzmann's chair (so cruelly evacuated a year before, 1906). Both Boltzmann and Mach were, as you know, just as much interested in philosophy, more especially in epistemology, as they were in physics; in fact all their later writing was pervaded by the epistemological ("erkenntnistheoretisch") out-look. Their views were not the same. But filled with a great admiration of the candid and incorruptible struggle for truth in both of them, we did not consider them irreconcilable. Boltzmann's idea consisted in forming absolutely clear, almost naïvely clear and detailed "pictures"—mainly in order to be *quite* sure of avoiding contradictory assumptions. Mach's ideal was the cautious synthesis of observational facts that can, if desired, be traced back till to the plain, crude sensual perception (pointer reading). He was most anxious not to contaminate this absolutely reliable timber with any other one of a more doubtful origin.

---

[1] *Editor's note:* A selection out of the letter he sent to Sir Arthur Eddington on March 22, 1940, was made by Schrödinger himself, and typed separately. It is this selection that is reproduced here.

However, we decided for ourselves, that these were just different methods of attack and that one was quite permitted to follow one or the other provided one did not lose sight of the important principles that were more strongly emphasized by the followers of the other one, respectively.

You easily imagine, Sir Arthur, that with these antecedents one cannot be very deeply impressed by a "brave new world" which, after having been taken in for a couple of decades altogether by one of these methods, finds itself suddenly let down by its alto[2] naïve application, then rediscovers the other one, which it proclaims as a new invention that has at last succeeded to uproot the old prejudices!

If I had had a dictaphone, I could quote you a conversation (with G. Kirsch) in which I tried to explain to him—we spoke about the terrible "jumps" in Bohr's orbits—that very probably the *place* of an electron within the atom had no meaning, because we had no means of observing it; that we may and must use pictures (Boltzmann), but with open eyes towards their limitations, which are given by what can be observed, because the ultimate aim of our pictures is only to serve as a scaffolding for our sensual perceptions (E. Mach). Maybe I am now embellishing the details—but the time is doubtless, by my remembering my room in the old institute, which we left in 1913.

. . . .

We never deal with just *one* atomic system (atom, electron, molecule). If, after having made a measurement, we make a second one (either a repetition or another measurement) we can never be *quite* sure that we perform it on the same individual atom. Indeed, apart from very special cases, the probability of it being the same is very low. The special cases are those where a particle is endowed with excessively high velocity (Wilson chamber).

I think, bearing this in mind, one can remove the silliness of present interpretation—the nonsense of the complete change in the wave function, alleged to be produced by a measurement—Procustation as you duly called it (so did I once in a little note).

---

[2] *Editor's note:* alto (or all-to) means "wholly." It is generally used with a verb (e.g., *alto dirtied* means *wholly dirtied;* see Oxford English Dictionary).

6 • WILLIAM JAMES LECTURES

This is the typescript of three William James lectures that Schrödinger was to deliver at Harvard University, c. 1954.[1]

I • First Lecture: Science, Philosophy and the Sensates

Being called upon as a scientist to lecture to a group of philosophers, I feel that the things I shall have to tell you in this course of lectures will, broadly speaking, fall under two categories. From the scientist you expect some information on science, more especially on his own science, physics, where he, maybe, speaks with some authority—as much as a human being at a certain epoch ever has in a certain realm of knowledge. This is to give you food, as it were, for your

---

[1] *Editor's note:* The story of these lectures is the following (see W. Moore, *Schrödinger, Life and Thought*, Cambridge University Press, 1989, p. 451): On December 15, 1952, Schrödinger was invited by W. V. O. Quine to deliver a series of William James lectures. He tentatively accepted to lecture for the fall term of 1954 and then began to prepare the texts he was to read in Harvard. But, due to a misunderstanding with Quine about the dates, he finally renounced. We are thus left with three unpublished texts which are likely to be only *part* of the William James lectures he would have delivered, but which are very polished and obviously ready for their intended use.
  The first and second lecture of this series were later considerably shortened and simplified to make Chapter 6 of Erwin Schrödinger's *Mind and Matter (Tarner lectures 1956, Cambridge)*, Cambridge University Press, 1958.
  As for the third lecture, it gives a very clear statement of Schrödinger's phenomenalistic conception of the "Thing" of everyday life. Other (less extensive) statements of this conception can be found in: E. Schrödinger, *Science and human temperament*, Norton and Co., 1935; E. Schrödinger, *Science and Humanism*, Cambridge University Press, 1951; E. Schrödinger, "What is an elementary particle?" Endeavour **9**, 109–116, 1950.

own philosophical thoughts. But as philosophers you may also be interested in *my* philosophical opinions, which I have formed from the sum total of my knowledge, mostly of the scientific kind,—formed it in my capacity as a human being. In respect of these things, I shall be speaking for myself only, and without any authority.—Now, of course, these two kinds of subject shall not be divided up into "part I" and "part II" of these lectures, they will be intimately interwoven. In giving me due credit—or withholding it,—you shall have to use your own discretion, according to the nature of the tenet.

What is philosophical thought—as opposed to scientific thought? Let me start from my own science. Physics is habitually divided into experimental and theoretical physics. The division is artificial. Physics consists in experimenting with nature, making careful observations, thinking about them, then planning and implementing new observations, suggested by the result of thinking, planning, experimenting and observing, in continual alternation. In this succession, thinking and planning is called theoretical physics, experimenting and observing is called experimental physics. The division is mostly caused by the fact, that the two kinds of activity require, each of them, such elaborate special training and skill, that they are seldom commanded by the same person—Ernest Rutherford and Enrico Fermi being rare exceptions.

The same state of affairs obtains in most other sciences, though not to the same extremity, because physics is the humblest, the lowest, the most basic, and therefore the most advanced of sciences.

I venture to maintain that philosophy bears to the sum total of sciences the same relation as the theoretical part of any particular science to its experimental or observational part. In attempting to unite the outcome of scientific research along the whole line to a picture of the external work, philosophy ought to acquire an important say in directing future research—not in detail but as regards the general attitude of how to conduct it.

I hasten to correct myself. Philosophy, in the quite general meaning of this word, is also concerned with things with which science—natural science—Naturwissenschaft—is never concerned. Philosophy also deals with *values*, with *ethics* and even with *esthetics*. These things are excluded from science, properly speaking, not by their nature, but by a simplifying convention which science has adopted and has not yet been able to discard, as I shall explain later in detail.

Thus philosophy has a much wider scope than I have circumscribed just before, indeed it has a very difficult task. For science has, as it were, monopolized the whole realm of experience and has arranged it in a scheme of thought which not only disregards ethical and esthetical values, but would be completely upset if they were admitted. Hence a philosopher who has decided on the sound principle of building on nothing but experience, because it is the only thing we have, finds the whole field occupied by science. And if he is naturally hesitant to interfere with the organisation of knowledge that science has achieved, the philosopher may be at a loss to find a foundation for the theory of values, including ethics.

I believe this to be the reason why ancient philosophy, after the truly important scientific advances of the Pre-socratics, including Democritus, turned more and more away from science, leaving it to itself, as it were. The great philosophical minds of the following centuries were keenly interested in the problems of ethics, where they may have felt the purely scientific world view more of a fetter than a help. Different attempts to bridge the gulf, which issued from Plato on the one side, from his great adversary Epicurus on the other, were not very fortunate; both men betrayed an equal lack of understanding of the scientific spirit.

Let us try and get at the root of this strange alienatedness of science from all kinds of value, ethical and esthetical and others, if others there are. Though the theory of values itself is not going to occupy us in these lectures, the reason why science is not only reticent on values but is anxiously on its guard not to have them introduced into its domain, is, so I believe, of primary interest, whenever scientific thought and philosophic thought is to be brought under a common scope, as is intended here. There must be a fundamental trait in the structure of scientific thought that excludes from its realm precisely what to the human mind in general appears to be of the highest relevance.

If you submit the scientific world view to a closer inspection, you will find that the picture it displays to us is a skeleton-like scheme, in which a lot more than just values is missing, things that are even nearer to us, but so to say on lower levels. If you ask a physicist what is his idea of yellow light, he will tell you that it is transversal electromagnetic waves of wavelengths in the neighbourhood of 590 millimicrons. If you ask him: But where does yellow come in? he'll say: In my picture not at all, but this kind of vibrations, when they hit the retina of a healthy eye, give the person whose eye it is, the sensation of yellow. On further inquiry you may hear, that different wavelengths produce different color-sensations, but not all do so, only those between about 800 and 400 mμ. To the physicist the infrared (> 800) and the ultraviolet (< 400) waves are much the same kind of phenomena as those in the region between 800 and 400 mμ, to which the eye is sensitive. How does this peculiar selection come about?— It is obviously an adaptation to the sun's radiation, which is strongest in this region of wave lengths but falls rapidly off sideways. Moreover the intrinsically brightest color-sensation, the yellow, is encountered at that place within that region where the sun's radiation exhibits its maximum, a true peak.

We may further ask, is radiation in the neighbourhood of wavelength 590 mμ the only one to produce the sensation of yellow? The answer is: not at all. If waves or 760 mμ, which by themselves produce the sensation of red, are mixed in a definite proportion with waves of 535 mμ, which by themselves produce the sensation of green, this mixture produces a yellow that is indistinguishable from the one produced by 590 mμ. Two adjacent fields, illuminated one by the mixture, the other by the single spectral light, look exactly alike, you cannot tell which is which. Could this be foretold from the wavelengths—is there a numerical connection with these physical, objective characteristics of the waves? No. Of course, the chart of all mixtures of this kind has been plotted empirically; it is called the color triangle. But it is not simply connected with the wavelengths. There is no general rule, that a mixture of two spectral lights matches one between them. E.g. a mixture of "red" and "blue" from the extremities of the spectrum gives "purple," which is not produced by any single spectra light. Moreover the said chart, the color triangle, varies slightly for one person to the other, and differs considerably for some persons, called anomalous trichromates, persons who are *not* color-blind.

The sensation of color cannot be accounted for by the physicist's objective picture of light-waves. Could the physiologist account for it, if he had fuller knowledge than he has, of the processes in the retina and the nervous processes set up by them in the optical nerve bundles and in the brain? I do not think so. We could at best attain to an objective knowledge of what nerve fibres are excited and in what proportion, nay even to know exactly the processes they produce in certain braincells—whenever our mind registers the sensation of yellow in a particular direction or domain of our field of vision. But even such intimate knowledge would not tell us anything about the sensation of color, more particularly of yellow in this direction—the same physiological processes might conceivably result in a sensation of sweet taste, or anything else. I mean to say simply this, that we may be sure there is no nervous process whose objective description includes the characteristic "yellow color" or "sweet taste," just as little as the objective description of an electromagnetic wave includes either of these characteristics.

The same holds for other sensations. It is quite interesting to compare the perception of color, which we have surveyed, with that of sound. It is normally conveyed to us by elastic waves of compression and dilatation, propagated in the air. Their wavelength—or to be more accurate their frequency—determines the pitch of the sound heard. (Nb.: The physiological relevance pertains to the frequency, not to the wavelength, also in the case of light, where however the two are virtually exact reciprocals of each other, since the velocity of propagation in empty space or in air does not vary perceptibly.) I need not tell you that the range of frequencies of "audible sound" is very different from that of "visible light," it ranges from about 12 or 16 per second to 20,000 or 30,000 per second, while those for light are of the order of several hundred (English) billions. The relative range, however, is much wider for sound, it embraces about 10 octaves (against hardly *one* for "visible light"), moreover it changes with the individual, especially with its age: the upper limit of pitch is regularly and considerably reduced as age advances. But the most striking fact about sound is, that a mixture of several distinct frequencies *never* combine to produce just *one* intermediate pitch as could be produced by *one* intermediate frequency. To a large extent the superposed pitches are perceived separately—though simultaneously—especially by highly musical persons. The admixture of many higher notes ("overtones") of various qualities and intensities results in what is called the varying timbre (German: Klangfarbe), by which we learn to distinguish a violin, a bugle, a church bell, a piano . . . even from a single note that is sounded. But even noises have their timbre, from which we may infer what is going on; and even my dog is familiar with the peculiar noise of the opening of a certain tin-box, out of which he occasionally receives a biscuit. In all this the ratios of the cooperating frequencies are all-important. If they are all changed in the same ratio, as on playing a gramophone record too slow or too fast, you still recognize what is going on. Yet some relevant distinctions depend on the absolute frequencies of certain components. If a gramophone record containing human voice is played too fast, the vowels change perceptibly, in particular the *a* as in "car" changes into that in "care."—A continuous range of frequencies is always disagreable, whether offered as a sequence, as by a siren or a howling cat, or simultaneously, which is difficult to implement, except perhaps by a host of sirens or a regiment of howling cats. This is again entirely different from light perception. All the colors which we normally perceive are produced by continuous mixtures; and a continuous gradation of tints, in a painting or in nature, are sometimes of great beauty.

The chief characteristics of sound perception are well understood by the mechanism of the ear, of which we have better and safer knowledge than of the chemistry of the retina. The principal organ is the *cochlea*, a coiled bony tube which resembles the shell of a certain type of sea-snails: a tiny winding staircase that gets narrower and narrower as it "ascends." In place of the steps (if we continue our simile) across the winding case elastic fibres are stretched, forming a membrane, the width of the membrane (or the length of the individual fibres) diminishing from the "bottom" to the "top." Thus, as the strings of a harp or a piano, the fibres of different length respond mechanically to oscillations of different frequency. To a definite frequency a definite small area of the membrane—not just one fibre—responds, to a higher frequency another area, where the fibres are shorter. These mechanical vibrations must set up, in *many different* nerve fibres, the well known nerve impulses that are propagated to certain regions of the cerebral cortex. I say in different nerve fibres, because we have the general knowledge, that the process of conduction is very much the same in all nerves and changes only with the intensity of excitation; this is rendered by the frequency of the pulses, which must, of course, not be confused with the frequency of sound in our case (the two have nothing to do with each other).

The picture is not as simple as we might wish it to be. Had a physicist constructed the ear, with a view of procuring to its owner the incredibly fine discrimination of pitch and timbre that he actually possesses, the physicist would have constructed it differently. But perhaps he would have come back from it. It would be simpler and nicer, if we could say that every single "string" across the cochlea answers only to one sharply defined frequency of the incoming vibration. This is not so. But why is it not so? Because the vibration of these "strings" are strongly damped. This, of necessity, broadens their range of resonance. Our physicist might have constructed them with as little damping as he could manage. This would have the terrible consequence, that the perception of a sound would not cease almost immediately when the producing wave ceases, but would last for some time, until the poorly damped resonator in the cochlea dies down. The discrimination of pitch would be obtained by sacrificing the discrimination in time between subsequent sounds. It is puzzling how the actual mechanism manages to reconcile both in a most consummate fashion.

I have gone here into some details, in order to make you feel, that neither the physicist's description, nor that of the physiologist contains any trait of the sensation of sound. Any description of this kind is bound to end with a sentence like: those nerve impulses are conducted to a certain portion of the brain, where they are registered as a sequence of sounds.—We can follow the pressure changes in the air as they produce vibrations of the ear-drum, we can see how its motion is transferred by a chain of tiny bones to another membrane, and eventually to parts of the membrane inside the cochlea, composed of fibres of varying length, described above. We may reach an understanding of how such a vibrating fibre sets up an electrical and chemical process of conduction in the nervous fibre with which it is in touch. We may follow this conduction to the cerebral cortex and we may even obtain some objective knowledge of some of the things that happen there. But nowhere shall we hit on this "registering as sound," which simply is not contained in our scientific picture, but is only in the mind of the person whose ear and brain we are speaking of.

We could discuss in similar manner the sensations of touch, of hot and cold, of smell and of taste. The latter two, the chemical senses as they are sometimes called, smell affording an examination of gaseous stuffs, taste of fluids, have this in common with the visual sensation, that to an infinite number of possible stimuli they respond with a restricted manifold of sensate qualities, in the case of taste: bitter, sweet, sour and salty and their peculiar mixtures. Smell is, I believe, more various, and particularly in certain animals it is much more refined than in man. What objective features of a physical or chemical stimulus modify the sensation noticeably, seems to vary greatly in the animal kingdom. *Bees*, for instance, are color-blind (dichromates) somewhat similar to the kind of vision found in about 3 or 4% of *men*. On the other hand, as von Frisch in Munich has found out not long ago, bees are peculiarly sensitive to traces of polarization of light, which aids their orientation with respect to the sun in a puzzlingly elaborate way. To a human being even completely polarized light is indistinguishable from ordinary, non-polarized. *Bats* have been found out as being sensible to extremely high frequency vibrations ("ultrasound") far beyond the upper human limit of audition; they produce it themselves, using it as a sort of "radar," to avoid obstacles.—The human sense of hot and cold exhibits the queer feature of "les extrêmes se touchent:" if we inadvertently touch a very cold object, we may for a moment believe that it is hot and has burnt our fingers.

• • •

2 • Second Lecture: The Technique of Measurement

Remember that we had started from the question, why in the scientific picture of the world, values are missing. We now believe to have found out, that something much more primordeal is absent from it, namely all our actual sensates, the primitive data of immediate perception. As soon as we have really understood this, it will not be difficult to get to higher levels, physical pain and physical delight, joy and sorrow, and possibly other features, intermediate between the sensates and the values, since on the one hand they are either directly attached to sensates or based on them, and on the other hand form the basis of values, though *not*—strangely enough—in the simple form that we deem valuable what promotes our own personal delight and joy and avoids pain and sorrow for us, devolving it on others.

But for the moment I wish to impress you with an exceedingly strange fact, an apparent paradox, which we shall try to solve. The scientific world picture does not contain the sensates, they are, unintentionally but quite systematically, removed from it. Yet this world picture is entirely based on sense perceptions, they are the material, the bricks of which it is built.—An electron is neither red nor blue nor any other color; the same holds for the proton, the nucleus of the hydrogen atom. But the union of the two in the atom of hydrogen, according to the physicist, produces electromagnetic radiation of a certain discrete array of wave lengths. The homogeneous constituents of this radiation, when separated by a prism or an optical grating, stimulate in an observer the sensations of red, green, blue, violet by the intermediary of certain physiological processes, whose general character is, however, sufficiently well known to assert, that they are not red or green or blue, in fact that the nervous elements in question display no color in virtue of their being stimulated; the white or grey they show whether stimulated or not is certainly insignificant in respect of the color sensation which, in the individual whose nerves they are, accompanies their excitation.

Yet our knowledge of the radiation of the hydrogen atom and the objective, physical properties of this radiation, originated from someone's observing those colored spectral lines in certain positions within the spectrum obtained from glowing hydrogen vapour. This procured the first knowledge, but by no means the complete knowledge. To achieve it, the elimination of the sensates has to set in at once, and is worth pursuing in this characteristic example. The color in itself tells you nothing about the wavelength, in fact we have seen before that e.g. a yellow line might conceivably be not "monochromatic" in the physicist's sense; it might be composed of many different wavelengths, if we did not know that the construction of our spectroscope excludes this. It gathers light of a definite wavelength at a definite position in the spectrum. The light appearing there has always exactly the same color from whatever source it stems. Even so the quality of the color sensation gives no direct clue whatsoever to infer the physical property, the wavelength, and that quite apart from the comparative poorness of our discrimination of hues, which would not satisfy the physicist. *A priori* the sensation of blue might conceivably be stimulated by long waves and that of red by short waves, instead of the other way round, as it is.

To complete our knowledge of the physical properties of the light coming from any source, a special kind of spectroscope has to be used; the decomposition is achieved by a diffraction grating. A prism would not do, because you do not know beforehand the angles under which it refracts the different wave lengths. They are different for prisms of different material. In fact, *a priori* you could not even tell that the more strongly deviated radiation is of shorter wavelength, as is actually the case.

The theory of the diffraction grating is much simpler. From the basic physical assumption about light—merely that it *is* a wave phenomenon—you can, if you have measured the number of the equidistant furrows of the grating per inch (usually of the order of many thousands), tell the exact angle of deviation for a given wavelength, and therefore, inversely, you can infer the wavelength from the "grating constant" and the angle of deviation.—In some cases, notably in the Zeeman effect and Stark effect, some of the spectral lines are polarized. To complete the physical description in this respect, in which the human eye is entirely insensitive, you put a polarizer (a Nicol prism) in the course of the beam, before decomposing it; by rotating the Nicol around its axis certain lines are extinguished or reduced to minimal brightness for certain orientations of the Nicol, which indicate the direction (orthogonal to the beam) of their total or partial polarization.

Once this whole technique is developed, it can be extended far beyond the visible region. The spectra lines of glowing vapours are by no means restricted to the visible region, which is not distinguished physically. The lines form long, theoretically infinite series. The wavelengths of each series are connected by a relatively simple mathematical law, peculiar to it, that holds uniformly throughout the series with no distinction of that part of the series that happens to lie in the visible region. These serial laws were first found empirically, but are now understood theoretically. Naturally, outside the visible region a photographic plate has to replace the eye. The wavelengths are inferred from pure measurements of lengths, first, once and for all, of the grating constant, that is the distance between neighbouring furrows (the reciprocal of the number of furrows per unit length), then by measuring the positions of the lines on the photographic plate, from which, together with the known dimensions of the apparatus the angles of deviation can be computed.

These are well known things, but I wish to stress two points of general importance, which apply to well-nigh every physical measurement.

The state of affairs on which I have enlarged here at some length is often described by saying that, as the technique of measuring is refined, the observer is gradually replaced by more and more elaborate apparatus. Now this is, certainly in the present case, not true; he is not *gradually* replaced, but from the outset. I tried to explain that the observer's colorful impression of the phenomenon vouchsafes not the slightest clue to its physical nature. The device of cutting a grating and measuring certain lengths and angles has to be introduced, before even the roughest qualitative knowledge of what we call the objective physical nature of the light and of its physical components can be obtained. And this is the relevant step. That the device is later on gradually refined, while remaining essentially always the same, is epistemologically unimportant, however great the improvement achieved.

The second point is that the observer is never entirely replaced by instruments; for if he were, he could obviously obtain no knowledge whatsoever. He must have constructed the instrument and, either while constructing it or after, he must have made careful measurements of its dimensions and checks on its moving parts—say a supporting arm turning around a conical pin and sliding along a circular scale of angles—in order to ascertain that the movement is exactly the intended one. True, for some of these measurements and check-ups the physicist might depend on the factory that has produced and delivered the instrument; still all this information goes back ultimately to the sense perceptions of some living person or persons, however cute devices may have been used to facilitate the labour. Finally the observer must, in using the instrument for his investigation, take readings on it, be they direct readings of angles or of distances, measured under the microscope, between spectra lines recorded on a photographic plate. Again, many helpful devices can facilitate this work, for instance photometric recording, across the late, of its transparency which yields a magnified diagramme on which the positions of the lines can be easily read. But they must be read! The observer's senses have to step in eventually. The most careful record, when not inspected, tells us nothing.

So we come back to this strange state of affairs. While the direct sensual perception of the phenomenon tells us nothing as to its objective physical nature (or what we usually call so) and has to be discarded from the outset as a source of information, yet the theoretical picture we obtain eventually, resides entirely on a complicated array of various informations, all obtained by direct sensual perception. It resides upon them, it is pieced together from them, yet it cannot really be said to contain them. In using the picture we usually forget about them—except in the quite general way we know, our idea of a lightwave is not an haphazard invention of a crank, it is based on experiment.

I was struck when I discovered for myself, that this state of affairs was clearly understood by the great Democritus in the 5th century B.C., who had no knowledge of any physical measuring devices remotely comparable to those I told you about (which belong to the simplest used in our time).

Galenus has preserved us a fragment (Diels fr. 125), in which Democritus introduces the intellect having an argument with the senses about what is "real." The former says: "Ostensibly there is color, ostensibly sweetness, ostensibly bitterness, actually only atoms and the void," to which the senses retort: "Poor intellect, do you hope to defeat us while from us you borrow your evidence? Your victory is your defeat."

To this discourse the great scholar of antiquity, Theodor Gomperz, remarks : one wonders what Democritus lets the intellect reply to this.—I presume, nothing. It would be too bad, if Galenus had quoted only a part of this interesting conversation. Anyhow, Gomperz, writing in the first decade of our century, still deems the correct answer to be about as follows: the information conveyed to us by the senses refers to "primary" and "secondary" properties of the things around us. The secondary ones are such features as red, blue, sweet, hot, cold, etc.; the primary ones refer to geometrical shape and arrangement, and to motion. The scientific world picture, so says Gomperz, is to be based on the latter only, because they are reliable, not on the former, which are not.

We do not accept this distinction today. The first to oppose it, so I am told, was Leibniz, but I have not looked up the reference given by Hermann Weyl (in his book on the philosophy of science). The following idea was wide-spread for a long time and may not be extinct today. We come to know the shapes and the relative positions and motions of the bodies in our environment, including our own body, by direct sense perception, mainly by the coordination of sight and touch. Science, as it were, sharpens our senses by the aid of instruments, and thus refines and extends this spatio-temporal geometrical and kinematical knowledge, extends it to the stars and nebulae, refines it to light-waves, atoms and molecules. Of the correctness of our views in these realms we cannot be as sure as we are in the case of the palpable and visible objects in our immediate surrounding. Hypotheses intervene and have sometimes to be replaced by better ones. But those who hold the view in question feel sure that some geometrical and kinematical structure of the world is a reality and that physical science aims at discovering this reality. Such sensual qualities, however, as red, sweet, hot . . . do not, so they say, form part of this reality, they and only they spring from the interaction of that "colorness and tasteless etc." external world with a peculiarly constituted organism and depend largely on its constitution.

This view has a great verisimilitude and deserves to be argued. The most obvious general argument against it is, that all sense perceptions depend on the interaction of something with our own body, also those which yield the spatio-temporal pictures, formed in every day life and in the scientific analysis. Moreover Einstein's theories of relativity have made it plain, that the arrangement of things in space and time is not the simple and obvious thing that it was formerly believed to be, but a rather sophisticated construct; and quantum theory has raised appreciable doubt, whether the construct adopted hitherto is really appropriate for accounting for all our experiences. We reserve these points for later consideration.

Let us for a moment return to the measurement of wavelengths of light. We find in that procedure traits that are peculiarly reminiscent of the view we have just dismissed on general grounds, traits moreover that are in common to most, if not all, physical measurements. Though one cannot, of course, dispense with the intercession of the observer's sensorium altogether, the device seems to be carefully chosen so as to make the results independent of who observes. If (for unfathomable reasons) a stone-blind person had to read the positions of the lines from the photometric diagramme, one would only have, by some chemical device, to transform it into Braille-script, to make this possible. We feel reasonably sure that with these aids the blind man would reach the same results as the seeing observer.

This is in striking contrast to observations that imply the judgement of sensate qualities as color, sound, hot or cold. I told you before that the color of "spectral yellow" can be matched by a suitable mixture of "spectral green" and "spectral red." Most normally sighted persons need very nearly the same ratio of "red/green" to obtain a perfect match. But there are persons who, while they produce and reproduce the match with the same accuracy, need a quantitatively different composition of the mixed light, e.g. 50% more green than a normal person. They are called anomalous trichromates. I happen to be one of them. If I put me to a color mixing apparatus, I shall easily produce "my" mixture; but not if you promised me a million dollars— not if my life depended on it—could I set the apparatus so that the two fields match for *you* (or most of you). Similarly a musical child, given a tuning fork of high pitch, but still well audible to all of us, might find it possible to set a Galton-whistle so as to give the next higher octave, while to an elderly person like me the whistling would be inaudible, so that he could not for love or money produce this setting by the ear for himself. (A Galton whistle is a small whistle whose pitch can be changed and read off a scale; it is used to determine the upper limit of audition for individual subjects.)—Even the same person can register very different impressions from the same object simultaneously. A very old little experiment to show it is this: keep for a minute or so the left hand in cold water, the right one in hot water, then put them both in a basin of tepid water that you have prepared; it will seem much warmer to the left than to the right hand.—When I was working in a dark-room and had to go out into the light to fetch something, I used the trick of blindfolding one eye. Thus it would preserve its adaptation and serve me after returning to the dark-room and removing the band as before, while the other one, being dazzled, saw the same surrounding pitch dark, as though it were actually blinded, at first a slightly uncomfortable feeling.

Our analysis of the measurement of the wavelength of light, which we use here as representative of a measurement in experimental physical science, has revealed two features of great importance that we shall have to discuss at length. They could be clinched, I dare say, by analysing any process of measuring used in physics or astronomy, because they are common to all of them, that is why they are so significant. One is, that the intervention of the observer's sensorium seems to be restricted to aver certain purely geometrical fact as lengths, angles, arcs, the location of certain marks in space (and perhaps in time); the other feature is, that the device always aims at making the result independent of the individuality of the observer, thus the same for all observers,—only then do we call it, in a sense, an objective result. These two things are hardly independent. But we must not prejudice their dependence; at any rate they are different, and each provokes ample discussion, so that we must deal with them one after the other. We take the first one first.

To aver geometrical facts is a very wide and loose expression which must be tightened for precision. How do we actually measure a length or an angle? We do not measure them, we compare them with standards. To measure a length we, in principle, have to put a measuring rod in contact with the object whose length we wish to ascertain. We let the zero of the rod coincide with one end of the object and observe that its other end coincides *at the same time* with a certain scale division of the rod. In this procedure the simultaneity of the two coincidences is of paramount importance. It is usually taken for granted, but sometimes needs careful attention. Two workers handling a long measuring tape or a surveyor's chain in laying out the plan of a house or garden, or of a tennis court, will communicate by brief shouts, which mean: Have you your end to the point?—Yes I have.—All right then, I take my reading and drive in my peg.

In laboratory work we use, as a rule, for measuring distances more sophisticated methods than just the juxtaposition of a measuring rod to the object. But when analysed they all amount to "simultaneous coincidences" or "pointer readings," as they have been called. For instance, the distance between two spectral lines, recorded on a photographic plate, can be measured with a suitably constructed microscope. Either the plate is fixed to a carrier which, by means of a transporting screw, can be moved at will in a direction orthogonal to the lines, moved with respect to the solid support on which the microscope and the plate-carrier are mounted; or alternatively, the plate remains fixed and the tube of the microscope is moved in the same manner. In both cases a drum, provided with a scale, must be turning with the screw past a fixed mark in immediate contact with the scale, so that the number of turnings and fractions thereof can be read on the scale. To measure the distance between two lines, you first move the plate (or the microscope, as the case may be) by mean of the transport-screw, until the cross-wire visible in the eye-piece *coincides* with the first spectral line; then you take a reading on the screw-scale, which means you aver the *coincidence* of the fixed mark with a certain scale division. Then you move on to the second line and do the same. The difference of the two readings[2] gives you the required distance, expressed by a thread-distance ("Ganghöhe") of the screw. If you know this in centimeters, an easy reduction yields the required distance in centimeters. If you don't, you have to replace the plate by a reliable standard scale of length and perform on it similar operations as before on the late; only the meaning is now just the reverse: you measure a length that you know, and thus you determine the thread-distance of the screw. In principle you must not even rely upon the screw being cut precisely uniformly. This means that you have to perform a great many operations on the standard measuring rod, to *calibrate* the screw, as this process is called, i.e. to ascertain the movement of the plate (or the tube) that is brought about by every individual turning of the transport screw.

This lengthy and perhaps a little tedious analysis of a standard example goes to show that here anyhow the observer's sensorium is charged with nothing but averring *simultaneous local coincidences*. May we be satisfied that this holds for *all* measuring devices in physical science? With just one or two small and not very important addition we may, so I believe. You know that physical quantities are ultimately reduced to the units of length, mass and time. We have dealt with length—at length. As regards mass, the most common device for measuring it, or rather for comparing a given mass with standard masses is the weighing scale. It indicates the equality of the two masses on the cups by the pointer pointing to zero when the beam comes to rest. So this case seems to be even a little simpler, since we have to aver only *one* coincidence (between pointer and scale-zero) not two simultaneous coincidences. This is quite a frequent occurrence. I believe it happens only when that moving part (of the instrument) on whose movement the coincidence depends, while remaining free to move, actually comes to rest. It seems to me, that in these cases the *permanence* of the one coincidence plays the part that the simultaneity of two coincidences has in the other cases.—To enter on more sophisticated methods of mass measurement would be very lengthy and would not, so I believe, bring to the fore any new relevant aspect.

Now what about time? Mostly the determination of a time

---

[2] The integral number of turnings are indicated by a second, rougher scale.

interval is reduced to the determination of a length or an angle, as by an ordinary clock or by the motion of the fixed stars. Yet a new feature may occasionally come in which is worth mentioning though it is perhaps not very relevant. An audible mark, f.i. a short "click," may replace the local coincidence. That is to say the observer avers the simultaneity of a local coincidence and a "click"—or even the simultaneity of two different "clicks." I shall give a few examples. A rough, but quite reasonable method of determining the velocity of sound is to observe the time between seeing and hearing some sudden event—say a hammer failing on an anvil—that happens at a distance, which must of course, be known. Of the two observations the first states the simultaneity of two local coincidences: the watch "showing a certain time" when the hammer touches the anvil (the finite velocity of light may be disregarded). But the second observation states the simultaneity of "a certain time on your watch" with the "bang" that you ear produced by the hammer. This is certainly in no way a local coincidence or a "pointer reading."

Now this is not a very scientific method. But I am told that, formerly at least, before the modern methods of electrical registration were available, astronomers while watching a star pass the meridian, would listen to the tickings of the clock, counting them and interpolating by estimation fractions of a second between the tick preceding and the tick following the moment of the passage. Here again an audible sign is one of the two events whose simultaneity, or approximate simultaneity has to be averred by observer. Experience shows, that in all these cases an element of subjectivity is brought in, the so-called personal equation of the observer. There is obviously a time-lag between the "actual happening" of an event and its "registration" in the observer's mind. This time-lag is not likely to be the same for the visual and for the audible event, and the difference of the two seems to vary with the observer and may have to be taken into account. (I might be objected, that this difference cannot be appreciable; for if it were we should be continually confused by hearing all the noises in our surrounding a little after or a little before seeing the events that cause them. But a long story could be told about the marvellous faculty we acquire in correcting automatically and unconsciously for considerable shortcomings in the functioning of our sense organs, and quite particularly in their coordination.)

In the last example—timing the passage of a star—it would not change the situation in principle, if the observer by pressing a button conveyed a mark to a moving registering tape which received also time marks, every second, from the clock. Indeed in this case the two "simultaneous" events are the coincidence between the picture of the star and the cross-wire in the eye-piece and the pressing of the button. The latter event is again not a visual coincidence but a voluntary act which must not be anticipated to—and actually will not—be exactly simultaneous with the "actual passing of the star." (Similar considerations apply to the very popular use of a stop-watch, for instance in timing a race; but here the two time-lags may cancel out, if the two observed events—start and finish—are of the same kind.) In order to avoid that, in timing an event, the observer be required to aver the simultaneity of heterogeneous subjective facts, as a visual coincidence, an audible sign, a voluntary act . . . , one has to let the event register itself automatically, with practically no time-lag. This is perfectly possible in some cases if a photographic film is used for the moving tape and a cathode-ray beam for making the marks. But it is certainly not always done, and it might prove extremely difficult in some cases, for instance in registering the passage of a faint star through the meridian. Thus I should say that the measurement of time, while it does not essentially imply a new principle, yet in actual practice frequently requires a slightly more complicated performance on the side of the observer.

• • •

3 • Third Lecture: The Part of the Human Mind

It has been held, that the part of the human mind in quantitative physical experimenting reduces, on careful epistemological analysis, to the averring of local coincidences or "pointer readings." I do not think this can be maintained if it is realized that every single experiment or observation a physicist or an astronomer carries out does not begin only in the laboratory or at the laboratory, but includes, from the epistemological point of view, all the manufactoring processes by which their instruments were made in the workshops and factories from the raw materials. Moreover the construction of an instrument in the workshop is based on previous laboratory work of the most various kind; to give just one example, the previous investigation of the mechanical and elastic properties of various materials tells us that steel is suitable for certain parts of the apparatus, while lead or india-rubber are not. If you follow this line of thought, considering also the optical and electrical devices used, you find that even the simplest quantitative laboratory experiment has an extended ancestry or pedigree of foregone laboratory and factory work on which it is based and which really form part and parcel or it.

Let us try and form an idea of the kind of geometrical facts that have to be ascertained or rather to be produced by factory work. A very primitive and fundamental requirement is to make a plane surface. Whatever attempt we have made to produce it, we must then test it. How can this be done? If we possess already another plane surface that is easy, we put the two in contact and turn and shift the one around on the other, and see whether they remain in contact without any wabbling occurring or gaps showing. What if we had no plane surface to start with? (This is after all the primordial situation.) Well, we would attempt to produce two, put them in contact and make the same tests. Supposing by grinding them against each other we have succeeded to satisfy the test, would this prove that both are plane? Indeed not; *one* might be spherically convex, the other concave, with the same radius of curvature. What next? One feels, one should like to use "the other side" of one of them—but this is impossible, because there is a material on the other side. So we manufacture a third workpiece carrying four protruding points, not in a straight line (a good choice would be, at the corner of an approximately equilateral triangle and in its centre). We grind the points, until the gadget can be applied anywhere to *one* of the surfaces, to be tested, without wabbling and without a gap (i.e. all four points must touch); if the same holds on the other surface, then both are planes.—It is useful to preserve the gadget, which is a primitive "spherometer," for using it in future to test the planeness of any surface.

A plane surface can serve to test the straightness of a (not too blunt) edge. We have to apply the plane to the edge and rotate the plane around the edge as far as possible. If no wabbling and no gaps occur, the edge is straight.—A plane surface can also serve to test whether the angle between two plane surfaces, ground on the same work-piece, is a right angle, but you have to test *three* such work-pieces simultaneously. Set two of them on the testing plane, with one of their planes respectively, and see whether the remaining two planes can also be brought into full contact. If you succeed in all three combinations of the three work-pieces, all three angles are right angles.

Clearly all this is only the stammering of the first few letters of a long, long alphabet, our instruments being, as it were, words and long sentences composed from the letters of this alphabet. Clearly we cannot scan the whole dictionary. The next very important step would be to consider the manufacture and the testing of surfaces of revolution, in particular a conical peg fitting tightly and exactly into a conical hollow so that one piece can perform a simple rotation around the other, but no other movement. Such parts are manufactured by means of a lathe, which, I dare say, is the most important instrument in a modern work-shop. But the lathe itself contains integrating parts of the same character on which its functioning depends. Where have they been manufactured? Of course, on another lathe. If we could pursue the pedigree in any particular case, we should doubtless meet with more and more primitive lathes, and as an early ancestor, several thousands or years ago, the potter's wheel. Anyhow the rotating parts in the earliest great inventions of the human genius, the cart-wheel, the potter's wheel and the mill-wheel, must have been carved from wood by hand and eye-measure and trial and error. The present perfection can only have been reached gradually after it had occurred to some-one that such a rotating part, set spinning by force, can be used to manufacture work-pieces of rotational symmetry. (The lathe was known in ancient time, the old Greek verb for working with lathe and chisel being . . . ; unfortunately handicraft and any kind of mechanical work was despised in antiquity as befitting slaves and "banauses;" thus we are less well informed on its progress and development than e.g. on the discovery of the properties of conic sections.)

In the same way all our present tools and machinery from

the simplest to the most elaborate form a complicatedly interacting "fauna" whose "species" descend from each other and thereby develop to greater consummateness, the higher perfection attained by one instrument enabling us to improve on one or many others. And by this I do not merely mean the evolution of ideas, of "blue-prints;" the descent of machines from each other is a real physical lineage, just as with living organisms, one being manufactured by means of the other. And the evolution is fairly slow. If you killed all horses alive now, there would be no more horses perhaps for ever, certainly for a very long time. If you destroyed all lathes on the earth, the recuperation would not take quite as long, but mainly because other machines with rotating parts could be adapted provisionally to serve as lathes. A very good and comparatively simple example of the "breeding of instruments" is afforded by the work of Henry Rowland (1848–1891), Professor at Johns Hopkins), when he started to produce those fine optical gratings that serve to analyse light, as I mentioned before. It is of paramount importance that the many thousands of furrows of such a grating should be as exactly equidistant as possible. The vital part of the apparatus by which the furrows are cut in the metal of a mirror is a transport screw that moves the cutting diamond from one position to the next, always by the same exceedingly small distance, of the order of a thousandth of a millimetre. The screw originally at Rowland's disposal was not very perfect. But it could be carefully calibrated, in the way mentioned before, and then, taking into account its deviation from uniformity, one used it to transport the chisel that cut another screw, which could if necessary be improved in the same fashion. The consummate instruments for cutting optical gratings which resulted from this "breeding of screws" are still, so I understand, a valuable possession of the Department of Physics in Johns Hopkins University, and by no means relegated to a historical collection.—This example goes to show very drastically that the experiment of a scientist begins in the work-shops and not only when he starts to take "pointer readings" at his instruments. Periodic errors in the positions of the furrows, as result most easily from imperfect screws, have the consequence, that perfectly homogeneous light is not collected in *one* line as it should, but to a small part in a few faint neighbouring companions of the main line, which are called "ghosts." This can deceive you into believing that the line is actually a multiplet; many lines really are, and these multiplets and their structure are usually of even greater interest to us than the precise wavelength of the main line. Hence those "ghosts" or fake multiplets can be very confusing. There are, of course, methods of revealing their nature; in particular different gratings will show different ghosts. But the safest line to follow is, to avoid them altogether, or at least to reduce their number by cutting the gratings to utmost perfection.

We have attained a rough survey of the kind of operations and tests that have to be performed in order to prepare and execute quantitative experiments, such as are deemed to procure us objective information on "the real world around us." These operations, since they include the manufactory of the instruments, do not consist merely in taking pointer readings. Yet it is true that they are concerned only with a very restricted domain of empirical statements, viz. with geometrical and kinematical relations, as opposed to statements on color, taste, hot and cold etc. We would, so I believe, find this result confirmed by a more thorough investigation, including the production of optical, electrical, radioactive apparatus as well as the great variety of devices used in the chemical laboratory. It is true that the qualities of sensates are frequently used as a guide in experimenting, particularly by the chemist: the color of a precipitate, the smell or taste of a drug, the cooling or warming up of a solution on adding a known chemical, are continually used as sources of information about a particular body to be tested. But observations of this kind do not enter into, and are no source of, the scientific picture the chemist eventually forms of the elements and compounds and their reactions—not even when he is practically interested in some of these sensates, as in manufacturing dyes or perfumes and coming to know by experience the effect that the substitution of a particular group is likely to have on the color or on the odour of a compound he is synthetising. His quantitative work depends on weighing, measuring volumes and concentrations, heat productions, time-rates of chemical reactions and whatnot, and resides ultimately on the same kind of geometrical and kinematical statements that we were speaking of before.

This whole situation has led to the very commonly adopted view that *something* is distributed in space in a definite arrangement and well defined order, the distribution or arrangement or order changing with time in a definite way, and this changing something is the objective reality at the back of "the world around us" including, of course, our own bodies (since your body forms part of my surrounding and mine part of yours). By calling it the objective reality people mean, that what they find out, be it by direct inspection or by more sophisticated methods, about the relative positions of parts of this something, about their geometrical configuration and its change in time, are hard, indubitable fact; or to put it more cautiously (since we may at times succumb to an error), that these findings are *about* hard, indubitable facts and are true or false according to whether they agree with them. What this *something* is, cannot be said; by calling it matter or field or whatnot, we just give it a name. The relevant point is that it is not supposed to have any other properties but geometrical configuration, changing in time according to certain "laws of nature." It is not in itself yellow or green, sweet or cold. If parts of it appears to us so, there is no hard, indubitable fact to make this judgement true or false.

This view is strongly supported by our analysis of actual experimental procedure, and it is attractively simple. It carries us comfortably a long way, indeed so long, that we may have forgotten its artificiality, when we meet the obstacles that it renders unsurmountable. So it is better to ask the naive but very pertinent question right away: how do red and yellow, sweet and hot come in at all? Once we have removed them from our "objective reality" we are at a desperate loss to restore them. We cannot remove them entirely, because they are *there*, we cannot argue them away. So we have to give them a living space, and we invent a new realm for them, the mind, saying that this is where they are, and forgetting that the earlier part of the story—all that we have been talking about till now—is also in the mind and nowhere else. But deeming it to be something else—objective reality—we run against the unanswerable question: how does matter act on mind, to produce in it the sensory qualities—and also how does mind act on matter, to move it at will? These questions cannot, so I believe, be answered in this form, and they owe their embarrassing form precisely to our having posited an objective reality which is a pure geometrical scheme of thought and deprived of everything real given by experience.

We shall return to this later. We have next to discuss the *second* feature that had clearly emerged from our analysis of a typical measuring device. This second feature is that our principal aim in quantitative experimenting is to adopt a procedure that makes the results independent of the individual observer. Unless we are fairly satisfied on this point, we do not attribute any significance to our findings. Our striving for *invariance* (as we may call it) is certainly not unconnected with the restriction of sensory averments to purely geometrical facts, but the former is doubtless the more fundamental thing. The motive for the said restriction is quite obviously our belief that with regard to the geometrical facts in question there can be no discrepancy in the judgement of two observers; and even the danger that fatigue or lack of attentiveness on the part of the observer adulterate his statements are reduced to a minimum. This is certainly true for what we called pointer-readings: the fact that the end of a narrow black line (the "pointer"), traced on one piece of metal, coincides exactly with the end of a narrow black line (a scale division) on another piece of metal in contact with the former, is so obvious, that we can hardly imagine two persons disagreeing about it. The worst that can happen is, that a skilled observer criticizes the setting of a less skilled one as being inaccurate, and corrects it, but the latter is not likely to disagree eventually. The other geometrical facts, mentioned before, as they occur in constructing an instrument—e.g. the requirement that a movable part of the apparatus turns exactly around an axis without wabbling—are less obvious; but the ultimate check-up on them in the laboratory can always be reduced to pointer-readings. For instance a suspected wabbling can be made very conspicuous by fixing a small mirror to the movable part and watching the picture of an incandescent wire thrown by the mirror onto a distant scale (either a projecting lens must be used, or the mirror must be slightly concave).

The aim of invariance, or independence of the individual observer, seems so natural that it does not appear to require an explanation. Still it is the fundamental thing. Why is it so natural? Because it is only the continuation of a behaviour that we have adopted from earliest babyhood, have developed to high perfection, and use every awake minute of our life for orientating ourselves in our daily surrounding.

The process is not very easy to describe appropriately not only on account of the abundant details it comprises, but mainly because in its most important early stages we are not aware of it; that is why I called it a behaviour, not a method (full awareness sets in only in the scientific stage). Now, to describe a thing means to make the listener aware of it. In the present case this creates the wrong impression as if we were speaking of a conscious process that involves a lot of reasoning, while it is a behaviour that we develop spontaneously and inadvertently.

To have a short name let us call it "forming invariants." It begins within the sensory complex of the individual, but very soon extends to forming mutual invariants, in common to the individuals that are in social contact. Any palpable object of interest in our environment appears in sight, which is the most important of our senses, under the most varying aspects of size, perspective distortion and illumination, which change continually. We gradually learn to disregard these changes or rather to unite the various aspects into the idea of a *thing* that does not change. Considering the enormous changes in the visual picture of the same "thing," this must be a very difficult task, though it is, of course, greatly aided by the senses of touch and sound. It is astonishing how quickly the baby does learn it with such emotionally stressed objects as the milk-bottle, the comforter, the rattle, and his own limbs, particularly the interesting toes. This useful forming of invariants, having been learnt at a tender age and practiced through life-time, has become such an ingrained habit, that we continually "see" features that we do not see. Looking at my table lamp I see that its socle is a black, shining square, though it actually appears as an irregular quadrangle, not altogether black, because one part reflects the bright sky; and the impression of shinyness is known to result mainly from the fact, that those reflections appear at slightly different places to my two eyes. Looking at a matchbox we see that it is a rectangular parallelopipedon, we would notice if it were crushed, even though the actual aspect of the intact box is rather "crushed" by perspective. The most amazing thing is our automatic "correction for distance," in other words our "seeing things in the right size," provided we have means to judge of the distance perceptionally, not merely by reason. We are so used to "seeing" familiar objects in the same size irrespective of distance, that we forget how strange this is. The heads of the people with us in a room all seem of about the same size, though the angular sizes may vary in the proportion 1:5 or 1:7; and they do not seem to change in size as people walk around the room. We only notice this automatic correction when in exceptional circumstances it produces an unexpected change of the apparent size of some object. If looking through a window at the house opposite or at a mountainous landscape, you step back a few yards, you will probably get the impression that the object outside grows, and that it diminishes in size if you approach the window again. What happens is that the window frame is psychologically corrected for distance, but the external object covers a greater range inside it, as you step back, it "expands" within the frame. The effect is strongest with objects that have no structure to betray their actual size, e.g. a rocky hill. Another example is the moon, which looks very much bigger near the horizon, just after rising or before setting, than when high up in the sky. The reason is this. We have, of course, no perception of the real distance of a celestial body, we project it to the vault of the sky. The latter does not give us the impression of a hemisphere, it appears considerably flattened, the hight seems considerably shorter than half the horizontal diameter. Thus when near the horizon, the moon is judged to be farther away and therefore to be bigger, if the angular size is the same. (It is not always the same, but this is not connected with its position in the sky; in fact the apparent change in size we are discussing here is much greater.) One may ask: why does the vault of the sky appear flattened? The reason that is usually given is that towards the horizon there are objects interposed, landmarks, which let you perceive the increase in distance, while towards the open sky there are none except perhaps clouds; in some cases they do make a difference, but as a rule they are not aligned and their structure gives no perceptual clue to their actual size.—The other day, when sitting in a standing bus at daytime, I saw through one of the windows, a few yards in front of me and viewed under a glancing angle, a big bright red neon-advertisement script which I took to be on a distant building on which it projected. This was "thoughtless" automatism, for the script was in mirror writing. Noticing this, I searched for, and discovered the original on the other side of the bus quite near to me. To my amazement I found the original very considerably smaller, so that for a second I doubted, did they really belong together. A repeated comparison left no doubt. So then I knew that they must be equal in size. But even this knowledge hardly removed the apparent discrepancy, caused by immediate perception, which located the mirror image on the distant building.

But let me return to the important general notion which I called forming invariants. The most fundamental invariants that we form at an early stage are the *things* in our surrounding, including our own body. We are, in this, greatly helped by the fact that many of these things are solid, almost rigid. For thus we are able to get the profound geometrical idea of movement: a body getting into different positions with respect to others without changing its shape and size. If all things in our environment were like molluscs, changing their shapes continuously, it would be hard to recognize an object as the same when it re-enters our ken, nay to form the idea of permanent things at all; the geometry of rigid movements (which is the simplest and most important of all geometries) would not be an early acquisition of childhood, but perhaps a late and very abstract construct of mathematicians. Helped by the principles of rigid geometry, the child soon forms the idea of true (as opposed to perspectivic) changes of shape, being relative movements of the parts of one thing. And the child has no difficulty in forming invariants also on non-rigid things such as a blanket, a cushion, and its own limbs. Particularly interesting among these non-rigid things is the flame, for instance that on the top of a candle. It has a certain permanence though it is certainly not rigid, it flickers. But you can also blow it out, then it disappears into nothingness. And you can light it again, then "it" is again there. The appearance of a flame out of nothingness, when you strike a match or press a cigarette-lighter, is an event of paramount and incessant interest to a child. One of the reasons, so I believe, is the paradox it offers, as soon as the idea of permanently existing things has been formed—the idea of conservation of matter, if you will allow me this expression here, which refers to the eventual scientific formulation of virtually the same aspect.

During the early years when the invariant notions of existing things are formed along with the notion that and how a thing can move and occupy other and other places, remaining yet the same thing, during these early years, I dare say, a child learns more of geometry and kinematics than in the ten subsequent years of schooling. It is relevant to stress, that these notions are formed by experience, by the experimental science of the baby and small child, and *not* to say: oh well, that is just only the way by which the child learns what the world really is like. The latter is true, but trivial. For by "what the world really is like" we mean the notion that we, the ordinary man or woman, have formed when we were small. That any small child going through similar experiences reaches the same aspect, is trivial and does not clinch the inevitability of this aspect. I range it with scientific constructs. This makes it liable to, and capable of, being subjected to revision and changed and improved, as all scientific theories are.

• About the Book

The text of this book is set in Trump Mediæval, a type family designed by Georg Trump in the late 1950s when Schrödinger was writing these essays. The math symbols are taken from the MathTime fonts designed to accompany Times Roman. Displayed heads are in Gill Sans, designed by Eric Gill in 1928 and still popular today.

The figures were redrawn from Schrödinger's originals, maintaining style and size wherever possible. Handwritten labels were replaced with type for readability.

The book was designed, typeset, and composed by Paul C. Anagnostopoulos of Windfall Software. The production was done with ZzTeX, a book composition package built on the TeX typesetting system by Donald E. Knuth.

The paperback cover photograph was taken by Prof. Dr. Wolfgang Pfaundler. Schrödinger's signature was recovered from his passport.